Introductory Digital I

Other Macmillan titles of related interest

W.A. Atherton, *From Compass to Computer*
B.R. Bannister and D.G. Whitehead, *Fundamentals of Modern Digital Systems*
J.C. Cluley, *Transistors for Microprocessor Systems*
Donard de Cogan, *Solid State Devices – A Quantum Physics Approach*
C.W. Davidson, *Transmission Lines for Communications, second edition*
M.E. Goodge, *Analog Electronics*
B.A. Gregory, *An Introduction to Electrical Instrumentation and Measurement Systems, second edition*
Paul A. Lynn, *An Introduction to the Analysis and Processing of Signals, third edition*
Noel M. Morris, *Electrical Circuit Analysis and Design*
Phil Picton, *Introduction to Neural Networks*
P. Silvester, *Electric Circuits*
L.A.A. Warnes, *Electronic and Electrical Engineering*
L.A.A. Warnes, *Electronic Materials*
B.W. Williams, *Power Electronics – Devices, Drivers, Applications and Passive Components*

Macmillan New Electronics Series
Series Editor: Paul A. Lynn

Graeme Awcock and Ray Thoms, *Applied Image Processing*
Rodney F.W. Coates, *Underwater Acoustic Systems*
M.D. Edwards, *Automatic Logic Synthesis Techniques for Digital Systems*
Peter J. Fish, *Electronic Noise and Low Noise Design*
W. Forsythe and R.M. Goodall, *Digital Control*
C.G. Guy, *Data Communications for Engineers*
Paul A. Lynn, *Digital Signals, Processors and Noise*
Paul A. Lynn, *Radar Systems*
R.C.V. Macario, *Cellular Radio – Principles and Design*
A.F. Murray and H.M. Reekie, *Integrated Circuit Design*
F.J. Owens, *Signal Processing of Speech*
Dennis N. Pim, *Television and Teletext*
M. Richharia, *Satellite Communications Systems*
M.J.N. Sibley, *Optical Communications, second edition*
P.M. Taylor, *Robotic Control*
G.S. Virk, *Digital Computer Control Systems*
Allan Waters, *Active Filter Design*

Introductory Digital Design

a_programmable_approach

Mark S. Nixon

Department of Electronics and Computer Science
University of Southampton

MACMILLAN

First published 1995 by
THE MACMILLAN PRESS LTD
Houndmills, Basingstoke, Hampshire RG21 2XS
and London
Companies and representatives
throughout the world

ISBN 0–333–61731–2

A catalogue record for this book is available from the British Library.

10   9   8   7   6   5   4   3   2   1
04 03 02 01 00 99 98 97 96 95

Typeset by
Richard Powell Editorial & Production Servs, Basingstoke, Hants RG22 4TX

Printed in Malaysia

Contents

Preface

Motivation

Why does anyone want to write yet another book on digital electronics? There are already many texts in this area, all of which follow a pretty similar format and are aimed at the hardware electronic engineer (and many of which are very large, excessively so in some cases). This book is different – it has an integrated hardware and software approach which is suited not only to electronic engineers but also to computer scientists and software engineers. At the same time, it does not omit electronic circuit analysis, concerning in particular the circuits used to implement digital electronics. It is shorter than many texts and aims to present the essentials of the subject, clarified by design examples.

The approach has developed from my experience in first year lecturing in digital electronics to aspiring electronic and software engineers. Electronic engineers cannot escape software any more (why they might want to, I do not understand) and neither can a computer scientist, who wants to grasp practical issues in computer system application, escape digital electronics. It is no surprise that the Institution of Electrical Engineers places much emphasis on the software engineering content of accredited electronics or electronic engineering courses. The software is introduced in the context of programmable devices, since these dominate much of modern digital electronic implementation.

Style

The book is formally structured, with diagrams integrated into the text. Any new terminology is introduced using italics and is explained directly thereafter. There are also comments which expand the material in the text. Each chapter concludes with further reading to give pointers for further study in specific areas. A selection of manufacturers' contact addresses is also given. The mixed logic convention for drawing digital electronic circuits is not used throughout the book but is introduced later in an appendix. My experience is that it can confuse matters greatly at an introductory level.

Design exercises predominate in later chapters and you can practise these, say, by implementation, or by varying the specification slightly and by checking that you achieve a workable and working solution. There are example questions at the end of each chapter and abbreviated solutions are given in a later appendix. The solutions can be obtained by anonymous FTP to ECS.SOTON.AC.UK where you LOGIN as anonymous, using the same password (reply anonymous to the Name and Password prompts), change directory to pub/digits (by cd pub/digits) where you will find a zipped file, digits.zip, containing source code for the design examples and the solutions to the programmable logic questions. Change the transmission type to binary (type bin) and obtain the file (type get digits.zip). This file needs to be unzipped so use PKUNZIP (use as PKUNZIP digits.zip [destination directory]).

If you have any comments concerning the material in this book or any suggested clarifications (dare I say errors?!) please email me at idd@uk.ac.soton.ecs. I promise that for good (new) suggestions I'll buy you a pint of beer!

Supplementary Texts

It is hoped that two further companion volumes to this book will be published. These would be:

1. *Volume 2: The Solutions Handbook.* This would contain fully worked solutions to all of the questions in this book.
2. *Volume 3: The Demonstration Handbook.* This would contain the design of a demonstration unit built and documented by Graham Jordan and the PALASM programs (and associated overheads) that I use (on the demonstrator) in my first-year digital-electronics lectures.

Both would be available at nominal cost. Volume 2 would be obtainable from Macmillan Press at the address given on page iv (you would ask for *Introductory Digital Design, Volume 2: The Solutions Handbook* by Mark S. Nixon). *Volume 3: The Demonstration Handbook* would be available from Mark S. Nixon at the Department of Electronics and Computer Science, The University, Southampton, SO17 1BJ, UK (ask for *Introductory Digital Design, Volume 3: The Demonstration Handbook* by myself and Graham C. Jordan) or email a request to idd@uk.ac.soton.ecs.

Acknowledgements

Many people have contributed either directly or indirectly to the material presented in this book. Many of my colleagues at Southampton have helped to clarify particular areas. I am particularly grateful to Adrian Pickering for his help and contributions, to John Carter for his observations and encouragement, and to Arthur Brunnschweiler for many discussions concerning electronics, digital or otherwise. I am grateful to the forbearance of my colleagues within the Vision Speech and Signal Processing research group and to Julian Field for his part in maintaining the computer systems and networks that I use, and to my students (past and present) who have made many helpful suggestions.

I am very grateful to AMD Inc who generously supplied a copy of PALASM free while I was writing this book, as they have done to many UK universities and colleges. Many PAL manufacturers, including AMD Inc, deserve applause for their consistently generous support to UK academe by providing software support free. AMD Inc even offered the PALASM suite to the UK academic community via the Rutherford Appleton Laboratory.

Dedication

I could dedicate this book to my son, whose energy and enthusiasm I covet, to my daughter, whose peace it shattered, or to Caz, for her constant support. I would even be happy to dedicate it to engineering as a topic for study and as a career, even in this climate where the "arts' seem to be increasingly popular. What a good idea!

Southampton, 1995   MSN

1 Introduction

1.1 Introducing digital electronics

A computer is a machine that executes instructions, called *software*, using electronic circuits, called *hardware*. The software is written using a *programming language*, which is translated so that it can run, or execute, in the electronic circuits. The hardware is made up from *digital electronics*. This book describes how you design the digital electronic circuits. These circuits are used more widely than just in computers. Around the home you will find them in many domestic appliances, even in many front door chimes.

This book describes the design of these circuits from a very basic form, and by the end you should be sufficiently equipped in knowledge to tackle the design of a computer system. The approach to this design uses an integrated software and hardware approach since this is the way many modern designers tackle the subject. This approach is a long way from the origins of digital electronics. It is a subject that has progressed rapidly with the awesome developments in electronic (and particularly microelectronic) technology. Its origins lie in studies by mathematicians in the nineteenth century. One may be familiar: though Lewis Carroll is perhaps most famous for *Alice in Wonderland* he was also (as Charles Dodgson) an eminent mathematician. However, it was George Boole who published *An Investigation into the Laws of Thought*, which was such a major contribution that Boolean logic algebra is named after him. This algebra had to await application in engineering until Claude Shannon implemented it in switching circuits using relays, which were the earliest form of switching device. Relay circuits progressed to valve circuits, but computers were simple, bulky and power-greedy. The next major breakthrough was the transistor, which miniaturised switching circuit implementation. When small numbers of switching circuits were combined (or integrated) in a single chip the technology was known as *small-scale integration* (SSI). As miniaturisation increased, so did the number of devices on a chip, and microelectronic technology matured through *medium-scale integration* (MSI) to *large-scale integration* (LSI) in the 1970s to *very-large-scale integration* (VLSI) in the early 1980s. The term VLSI still applies, though *ultra-large-scale integration* (ULSI) finds some favour.

Design techniques have developed concurrently with these hardware developments and design techniques have even been motivated by technological development. Some design techniques have little relevance now, having been developed concurrently with, say, MSI or LSI technology. The main developments in recent years have included programmable systems, which require software specification of a device's function. This is why this book uses an integrated software and hardware approach.

1.2 Organisation of this book

The book starts with the study of combinational logic. This is the basis of the subject and several alternative presentations can be found in the selection of texts referenced at the end of Chapter 2. Chapter 2 shows how basic *logic circuits* can be designed to implement digital functions. The major difference in the approach in this book is the use of software to present combinational logic systems.

Logic circuits are implemented using *electronic circuits*, which are described in Chapter 3. Computer science students can study these circuits to appreciate the advantages of particular technologies, since these confer advantages to the computer systems that they program and interface. Electronics students should note that the treatment of microelectronic technology here concerns switching circuits only, though these are just a branch of a much wider subject. Digital circuits are implemented using *logic technologies*; these continue to evolve and this presentation is a snapshot of those technologies predominating at the time of writing.

A major part of digital electronics is *sequential system design*, which concerns digital circuits that go through a sequence of specified conditions, or states, as in a computer program. This is presented in Chapter 4, again using software as part of the circuits' descriptions.

Programmable logic, which is digital electronic hardware whose connections are specified by software in *programmable logic devices*, is introduced in Chapter 5, which surveys the major programmable technologies that are currently available and concentrates on *programmable array logic (PAL)* which has rapidly become the most popular means for digital circuit implementation. Several design examples, implemented using PAL technology, serve to reinforce the advantages that a programmable approach imbues.

A number of textbooks on digital electronics actually start with *number systems* and their arithmetic, concentrating on the *binary system* in particular, to introduce digital electronics as a vehicle for implementing binary arithmetic. This subject is treated later in this book, in Chapter 6, in part to allow for inclusion of design examples concentrating on circuit design. This allows computer scientists, in particular, to appreciate exactly how the circuits at the heart of a computer are designed. The design examples concentrate on comparing the advantages of different approaches, contrasting results achieved with those of commercial designs.

This is followed by a study of *analog-to-digital (A/D)* and *digital-to-analog (D/A)* conversion in Chapter 7. This concerns how a digital (computer) system can talk to the outside (analog) world. This topic is often omitted from textbooks on digital design, which is surprising because many computer programmers are interested in how the signals that they are processing are acquired. Also, A/D and D/A conversion offers a rich topic for the study and design of combinational and sequential logic, in particular sequential logic.

Finally, Chapter 8 on *hybrid circuits and interfacing* covers some basic issues in practical digital electronic circuits. Digital electronic circuits need external control and need to interface to external circuits. These circuits are rarely completely digital, and an introduction to hybrid circuits serves to illustrate not only how you can build complete systems, but also how you can solve some of the difficulties in their implementation.

1.3 Preliminaries

A number of assumptions have been made concerning prior knowledge, and the most important of these are reviewed briefly here. Though binary arithmetic is studied in Chapter 6, earlier chapters do assume that you know what binary numbers are. The representation of a binary *integer* (without any fractional part) uses symbols 0 and 1 as *binary digits* or *bits* in the number. This can be compared with decimal arithmetic which has ten symbols, from 0 to 9, which are used for the *decimal digits*. In decimal arithmetic the four-digit integer 8056 is actually

$$8 \times 1000 + 0 \times 100 + 5 \times 10 + 6 = 8 \times 10^3 + 0 \times 10^2 + 5 \times 10^1 + 6 \times 10^0$$

Each decimal digit is then multiplied by a factor of 10. This factor corresponds to the position or significance of the digit, which increases from the right (in this case 6 is the least significant digit) to the left (and 8 is the most significant digit). The basis of the decimal system is 10, and that is why we have ten symbols and a factor of ten multiplying each.

The basis of the binary system is 2 and we have two symbols, 0 and 1. A binary number 1011 is actually

$$1 \times 2^3 + 0 \times 2^2 + 1 \times 2^1 + 1 \times 2^0 = 1 \times 8 + 0 \times 4 + 1 \times 2 + 1$$

Since the base is smaller, and we have fewer symbols, the numbers need more digits. A binary count sequence compared with its decimal equivalent is

Decimal	Binary	Decimal	Binary	Decimal	Binary	Decimal	Binary
0	0 0 0 0	4	0 1 0 0	8	1 0 0 0	12	1 1 0 0
1	0 0 0 1	5	0 1 0 1	9	1 0 0 1	13	1 1 0 1
2	0 0 1 0	6	0 1 1 0	10	1 0 1 0	14	1 1 1 0
3	0 0 1 1	7	0 1 1 1	11	1 0 1 1	15	1 1 1 1

The main advantage of the binary system is that it is very convenient for digital circuits in general and computers in particular. It is covered in detail later, in Chapter 6, where other number systems are introduced.

Electronic circuit theory is the mathematical theory describing the operation of electronic circuits. Here follows a very basic introduction. It concerns in particular the relation between voltage (or potential) and current (the flow of charge or electrons). Using water as an anology, we can consider voltage as water pressure and current as the rate of water flow. Water flows from the supply to the destination, induced by the pressure difference between the two. Similarly, current in electronic circuits is related to voltage potential. Electronic current, denoted i, is shown by an arrow to flow from the positive power supply to the negative power supply and is depicted as opposite to the flow of electrons, which are negative charges and thus attracted to the positive voltage supply.

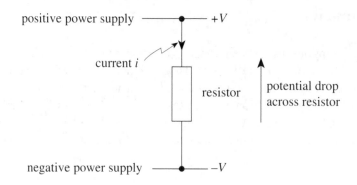

The relation between voltage, resistance and current is given by Ohm's law as

$$\text{voltage } (V) = \text{current } (I) \times \text{resistance } (R)$$

This implies that for a constant resistance, if we increase the voltage drop, then the current increases as well; by analogy greater water pressure increases the rate of flow. Also, for a constant voltage supply then an increase in resistance implies a reduction in current.

These concepts of binary numbers and electronic circuit theory should serve as a basic introduction to the remainder of this book. We shall start at the beginning of digital circuit design, namely combinational circuit design, which is the basis of the whole subject.

2 Combinational Logic Design

'Contrariwise', continued Tweedledee, 'if it was so, it might be, and if it were, it would be: but as it ain't, it ain't. That's logic.'

Lewis Carroll, *Through the Looking Glass*

2.1 Combinational logic/Boolean algebra

We shall use Boolean algebra to develop digital electronics. In its broader form it is known as discrete mathematics which, among other subjects, has wide application and implication in software engineering. This introduction is functional and aims to serve only as an introduction to digital electronics. For a better treatment (albeit one that sometimes seems pedantic to an engineer) you should consult one of the texts cited at the end of the chapter (p. 45).

One definition of *logic* is as a chain or science of reasoning. For Boolean algebra we are concerned only with truth or falsity. We are concerned with propositions that can have either of two values: *FALSE* or *TRUE*. A *proposition* is a variable or fact. For example, the proposition 'The cup is not cracked' is TRUE if the cup is indeed whole, but FALSE if the cup is cracked. A *statement* is formed by joining propositions; the overall validity of the statement depends on the validity of each proposition from which it is formed. Consider for example the statement 'A good cup is unbroken and glazed.' (This will later be formulated in terms of industrial quality control.) If the proposition 'The cup is unbroken' is TRUE and the proposition 'The cup is glazed' is TRUE then the statement 'The cup is good' is TRUE, since it is neither broken nor unglazed. This can be summarised using a *truth table*, which shows the validity of a statement according to the validity of its constituent propositions.

Note here that the cup can only be broken or not – just one of two values. It cannot be slightly broken. This is in accordance with Boolean algebra, where variables can have two states only.

A truth table for a statement made from two propositions is a list of the validity according to the four possible sets of the two propositions. For propositions A and B the four sets or combinations are:

(a) both FALSE; (b) A FALSE, B TRUE; (c) A TRUE, B FALSE; and (d) both TRUE.

Using FALSE = F and TRUE = T, the truth table for a statement which is a function of these two variables, $f(A, B)$, is

A	B	$f(A, B)$
F	F	?
F	T	?
T	F	?
T	T	?

The question then remains of determining whether the output is TRUE or FALSE for each input combination. Take the statement 'A good cup is unbroken and glazed.' This can be broken into a statement 'The cup is good' and two propositions, 'The cup is unbroken' and 'The cup is glazed.' The truth table then indicates whether or not the cup is good according to whether it is broken (or not) or glazed.

A = proposition 1, 'The cup is unbroken' (TRUE = unbroken, FALSE = broken)
B = proposition 2, 'The cup is glazed' (TRUE = glazed, FALSE = unglazed)
$f(A,B)$ = statement, 'The cup is good'

Propositions		Statement	
A	B	$f(A, B)$	
F	F	F	The cup is broken and unglazed; it is not a good cup
F	T	F	The cup is glazed but no good because it is cracked
T	F	F	The cup is unbroken but it has no glaze and is thus no good
T	T	T	The cup is good because it is neither broken nor unglazed

2.2 Logic functions

What we have actually defined is a way of linking variables to form a result. As in traditional mathematics, the variables are linked by a function. The function that we have seen illustrated is actually the *AND* function, which is perhaps unsurprising since the propositions were connected using the word 'and'. The truth table for the AND function is then

A	B	A AND B
F	F	F
F	T	F
T	F	F
T	T	T

This shows that the AND function is identical to the way that we use the word 'and' in language. The function A AND B is only TRUE when A is TRUE and B is TRUE, otherwise it is FALSE.

There is also a logical *OR* function, which again follows the way we use the word in language. Consider 'I will watch television if I have nothing else to do or there is a programme I really want to see'. Given the statement 'I will watch television', this will be TRUE if either or both of the propositions 'There is a programme I really want to see' and 'I have nothing else to do' are TRUE. The OR function can be summarised by its truth table:

A	B	A OR B
F	F	F
F	T	T
T	F	T
T	T	T

The AND and OR functions are basic logic functions. The last member of the basic set is called *NOT*. Its function is implicit in its name; it provides a logical **complement** or *inversion*. NOT TRUE is FALSE; NOT FALSE is TRUE, which gives a truth table:

A	NOT A
F	T
T	F

These functions are actually mathematical functions, and there is a **logic algebra** formulated around them. This is the basis of discrete mathematics. The aim here has been to introduce combinational logic as part of digital electronics; we shall now move to software specification and to logic circuits, and then define the logic algebra later in the context of combinational logic design.

2.3 Combinational logic and computer software

We can program computers to implement logic in *software*, a (computer) *programming language*. Software comprises a set of *instructions*. The simplest instruction is *assignment*, where we assign a value to a variable, in our case a Boolean value (TRUE or FALSE), to a logic variable. This can be of the form

```
variable_A = TRUE      variable_C = variable_A OR variable_B
```

When these instructions are executed (performed), in the first a TRUE value will be assigned to the variable_A, and in the second, variable_C is a function of two variables and will become TRUE when variable_A or variable_B is TRUE. This software is first created using an *editor*, and is then turned into a form suitable for execution within the computer using a *compiler*. The compiler will interpret the word OR to imply that we want the computer to execute the OR function at that point. This is then a *keyword* or *reserved word* for the compiler, and this fact is made clear to the reader by using capitals in the edited software or *code*. If the program is written incorrectly and the *syntax* is wrong then the compiler will fail to produce an executable version and will usually tell you where and why it failed. Words that are not keywords are indicated by lower-case type, and so lower-case is often used for variables.

> *This is a simplified description of a computer language – consider it as BASIC (a simple and popular language) for digital electronics. C^{++} buffs should skip this section!*

Statements are usually separated by *punctuation marks*, so we use spaces to separate keywords and variables so that the compiler interprets the software correctly. We cannot use spaces within variable names, so we use an underline to replace the space if we need to connect words; this is why variable_A is not written as variable A. We also use brackets to denote separation of statements. These are best sprinkled liberally in software to ensure that the compiler understands software commands correctly. Note that

```
variable_1 = variable_2 AND variable_3 OR variable_4
```

could be interpreted to give two different functions:

```
(a)   variable_1 = (variable_2 AND variable_3) OR variable_4
(b)   variable_1 = variable_2 AND (variable_3 OR variable_4)
```

so we should include brackets to ensure that the statement is specified unambiguously.

> *Syntax is used here the same way as it is in language: to specify grammatical arrangement. Software syntax defines how reserved words connect variables and the punctuation specifies their interpretation.*

The assignments that we make are often conditional; we use an IF statement that executes an assignment if the condition specified in the IF statement is TRUE, and its syntax is

```
IF condition is met THEN do something
```

We often need to specify what to do if the condition specified in the IF statement is not TRUE (i.e. the condition is not met). We do this by including an ELSE statement:

```
IF condition is met THEN do something ELSE do something different
```

e.g.

```
IF need_sugar    THEN add_sugar=TRUE ;If we need sugar then add it
                 ELSE add_sugar=FALSE    ;otherwise we do not
```

Here the semicolons precede *comments* on a program that we use to clarify the program's function, the text following the semicolon until the end of the line does not contribute to the compiled code. When this instruction is executed, if the current value of need_sugar is TRUE then add_sugar becomes TRUE, otherwise need_sugar is FALSE and so add_sugar becomes FALSE. This statement could clearly be simplified to

```
need_sugar = add_sugar
```

However, the condition supplied to the IF statement could be computed as a logic function of a number of logical variables. Also, the THEN and ELSE statements could have a sequence of instructions rather than a single logical assignment. We group commands, or **bracket** them, using BEGIN and END statements; for example,

```
IF ((need_sugar) AND (have_sugar))
                THEN BEGIN
                     add_sugar=TRUE   ;add sugar to the cup
                     used_sugar=TRUE ;adding sugar uses it up
                     END
                ELSE BEGIN
                     add_sugar=FALSE   ;either we don't need
                        ;sugar or we don't have any
                     used_sugar=FALSE ;so we won't use it up
                     END
```

We can also **nest** multiple IF statements to extend them to more than one level, e.g.:

```
IF condition_1 THEN
                IF condition_2   THEN function_1
                                 ELSE function_2
                ELSE function_3
```

This is executing a group of statements that can be expressed separately as

```
BEGIN
IF (condition_1 AND condition_2) THEN function_1
IF (condition_1 AND NOT(condition_2)) THEN function_2
IF (NOT(condition_1) AND condition_2) THEN function_3
IF (NOT(condition_1) AND NOT(condition_2)) THEN function_3
END
```

Function_3 is executed when condition_1 is FALSE whatever the value of condition_2. This is a rather prolix description, so we nest multiple IF statements to reduce complexity and to make the resulting code more readable. A multiple IF statement can be further simplified by using a CASE statement, where we look up the appropriate action from a list (or table) of possible actions. The particular value we choose is specified by the value of the conditions in the CASE statement. Its syntax for a single input or condition is

```
CASE (input)
BEGIN
   0: function_1 ;execute function_1 if input=0
   1: function_2 ;execute function_2 if input=1
END
```

This executes either of two functions according to the value of the input. The input can be either FALSE (0) or TRUE (1) and the first value in the table is selected if the input (the condition of the CASE statement) is FALSE and the second if it is TRUE. The colon is a punctuation mark in the case statement and the BEGIN and END statements are used to surround the assignments made in the CASE statement. This is then equivalent to

```
IF NOT(input)  THEN function_1
                ELSE function_2
```

The value of the CASE statement lies more in the simplification of multiple IF statements, e.g.

```
CASE (input_1,input_2)
BEGIN
   00: function_1
   01: function_2
   10: function_3
   11: function_4
END
```

This is an ordered list, giving a table for all possible values of the two conditionals (input_1 and input_2). Its function is equivalent to

```
IF (NOT(input_1) AND NOT(input_2)) THEN function_1
IF (NOT(input_1) AND input_2) THEN function_2
IF (input_1 AND NOT(input_2)) THEN function_3
IF (input_1 AND input_2) THEN function_4
```

The functions in the CASE statements are then written using the input ordering of a truth table and the output function placed at the appropriate position in the table. For some compilers the input ordering is **implicit** in the syntax and you do not specify which input combinations lead to chosen output functions; this is determined by the position of the function in the table. We shall be using software first to describe logic circuits and then to design them.

2.4 Combinational logic implementation

We have so far considered Boolean algebra as a formalism to demonstrate logical relations. Consider for example the problem of the cup which when it is on a production line has a natural system implementation: a quality control system to remove the cup from a conveyor belt.

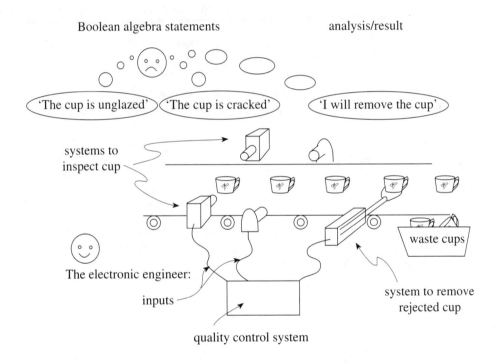

In your studies, away from digital electronics, you will doubtless learn how to design systems to measure whether the cup is cracked (can we use ultrasound? or computer vision?) or whether it is glazed (can we measure reflectance?), and systems to remove the cup from the production line. Before that, though, we need to design a system which can interpret the signals from the sensors, to combine them together to provide a signal to remove the cup. This is combinational logic, the implementation of Boolean algebra in switching circuits. Logical variables, *inputs*, are connected by logical functions, *gates*, to give a logical result, an *output*. A gate is an electronic circuit that implements a logic function via a switching circuit.

2.4.1   Combinational logic levels

We need signals to represent true and false. These are necessarily electronic signals, since we shall implement them in electronic circuits. We could use current or voltage, and both are indeed used in logic circuits. At first, though, we shall use *level logic*, voltage levels to denote TRUE and FALSE:

$$\text{TRUE} = \text{'1'} = +5\,\text{V} \quad \text{and} \quad \text{FALSE} = \text{'0'} = 0\,\text{V}$$

We tend to use the inverted commas to denote a logical '1'. This is because we are using a binary system with just two values and this is just one way of indicating it.

> *The use of an explicit level for TRUE and FALSE would be very difficult to achieve. We actually use a range of levels (mainly to reduce the possibility of error).*

The use of a high level to indicate a '1' and a low level to represent a '0' is called *positive logic*. You will meet this often in *data books*, which define how circuits operate.

2.4.2   Combinational logic gates

(a) AND

The AND gate is a circuit implementing the logical AND function. The AND function is TRUE when both propositions are TRUE; the output of an AND gate is '1' when both inputs are '1'. If either input is a '0' then the output is '0'.

Truth table

A	B	$A \cdot B$
0	0	0
0	1	0
1	0	0
1	1	1

Symbol

The raised dot symbolises the AND function – it looks like multiplication and in a loose way we can now think of the AND function as to multiply. In terms of a switching circuit we can draw the AND function as the collection of electrical switches to switch on a light. The switches are normally open (which represents '0') and close when pressed (representing '1').

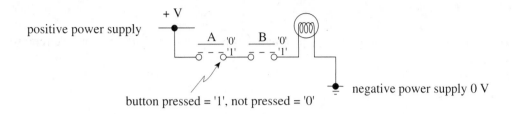

positive power supply

button pressed = '1', not pressed = '0'

In the AND function we need both inputs to be a '1' before the output is a '1'; here we need both buttons to be pressed before the light is switched on.

(b) OR

Truth table

Symbol

A	B	A + B
0	0	0
0	1	1
1	0	1
1	1	1

OR is symbolised by addition, which forms a useful memoriser. It is a loose mnemonic, since $1 + 1 = 1$! (We shall study arithmetic systems later.) The switching circuit for OR is given by two switches in parallel. If we press either or both buttons the lamp will light.

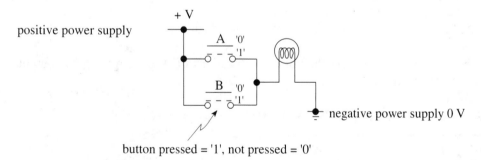

button pressed = '1', not pressed = '0'

Note that diodes can be used to replace the switches in the OR circuit. If the diodes' cathodes are used as logic inputs, and their anodes are connected to the lamp then we effectively form a diode OR gate, see Section 3.2.1.

(c) NOT

The NOT function is symbolised by placing a bar across a logic variable; thus $\overline{A}$ represents NOT A:

Truth table

Symbol

A	$\overline{A}$
0	1
1	0

$A\!-\!\triangleright\!\circ\!-\!\overline{A}$

2.4.3   Circuits and truth tables for more than two inputs

Truth tables and gates have so far been designed in terms of two inputs only. The truth table is merely a list showing the value of an output (or statement) for a combination of inputs (propositions). It can easily be extended to handle more than two input variables. Consider a function 'A cup is good if it has a handle, it is unbroken and it is glazed'. This function is TRUE when all three propositions are TRUE. In terms of a truth table, we need to tabulate whether or not the output is '1' (TRUE) according to the status of the three propositions. Each of the propositions can have either of two values and so there are $2 \times 2 \times 2$ (i.e. 8) combinations. The truth table is then formed from the variables or propositions. The propositions are

A = the cup has a handle ('1' = TRUE = it has a handle)
B = the cup is unbroken ('1' = TRUE = unbroken)
C = the cup is glazed ('1' = TRUE = glazed)
$f(A, B, C)$ = the cup is good ('1' = TRUE = good cup)

The truth table then covers the eight combinations of A, B, C giving the output value $f(A, B, C)$ for each combination:

A	B	C	$f(A, B, C) = A \cdot B \cdot C$
0	0	0	0
0	0	1	0
0	1	0	0
0	1	1	0
1	0	0	0
1	0	1	0
1	1	0	0
1	1	1	1

3-input AND gate symbol

A—
B— $A \cdot B \cdot C$
C—

The function is TRUE only when all inputs are TRUE, and therefore describes a 3-input AND function. It is implemented using a 3-input AND gate which provides, as output, a '1' (5 V) when all inputs are '1' (5 V) and its symbol is the AND symbol with three input connections.

> *There are two ways to construct a truth table. One is to list all input combinations by counting up in binary. The other way is to take an input and as a column write it down as 01010101, take the next and write its values as 00110011, and the last as 00001111. Note that if you go too far, you will just repeat the truth table.*

There is also a 3-input OR function which is TRUE if one or more inputs is TRUE. It is implemented using a 3-input OR gate which provides an output '1' (5 V) when any of the inputs is '1' (5 V), with truth table:

A	B	C	A + B + C
0	0	0	0
0	0	1	1
0	1	0	1
0	1	1	1
1	0	0	1
1	0	1	1
1	1	0	1
1	1	1	1

3-input OR gate symbol

This can of course be extended into truth tables that describe circuits with four inputs or more. Note that for n inputs, there are 2^n combinations. A truth table for 10 inputs tabulates the value of the output for 2^{10} (i.e. 1024) combinations and is rather large. Not all circuits reduce to a simple gate implementation such as a 3-input AND gate; most usually reduce to a collection of different gates.

2.4.4 Don't-care states

When implementing designs there are sometimes input combinations where the status of one input cannot affect the output. This may be either because it has no effect in that combination, or because it physically cannot occur. These are called *don't care states* and they are usually denoted using an X in the truth table; this signifies that the signal can be '0' or '1' and its value has no consequence in that state.

For a don't-care input, the following truth tables are equivalent:

A B	f(A, B)		A B	f(A, B)
1 X	1	≡	1 0	1
			1 1	1

2.4.5   *Further logic gates*

Two other important gates are formed by combining NOT with AND and OR. The bubble on the inverter symbol actually symbolises inversion, so the combination of NOT and AND, called *NAND*, is given by the inversion of AND (i.e. NOT AND) and is symbolised by the NOT bar across the AND function, and by the inversion bubble attached to the AND symbol:

Truth table

A	B	A · B	$\overline{A \cdot B}$
0	0	0	1
0	1	0	1
1	0	0	1
1	1	1	0

Symbol

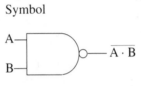

A switching circuit can again be constructed, but in this case by using switches which are normally closed (representing '0' here) and open when pressed (representing a '1'):

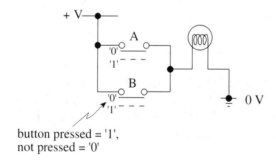

button pressed = '1',
not pressed = '0'

Only when both buttons are pressed does the lamp switch off. If only one or neither button is pressed, the lamp lights.

The combination of NOT and OR gives NOT (OR) called *NOR*. The truth table and symbol are then

A	B	A + B	$\overline{A+B}$
0	0	0	1
0	1	1	0
1	0	1	0
1	1	1	0

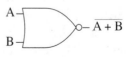

with a switching circuit again using normally closed switches:

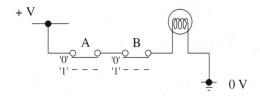

Here the lamp goes off if either button is pressed.

> *The AND circuit is formed by a series of switches, whereas for an OR function they are in parallel. For NAND the switches are in parallel and for NOR in series, but with the normally closed switches rather than normally open switches. This illustrates that the NAND gate can be implemented as an OR gate with inverted inputs; it is often drawn this way, as part of a circuit-drawing convention.*

Other symbols

Other logic symbols are used, for example those defined according to British Standard BS3939:1985 and the Institute of Electrical and Electronic Engineers Standard IEEE/ANSI 91:1984); these are given in more detail in Appendix 1 on drawing logic circuits.

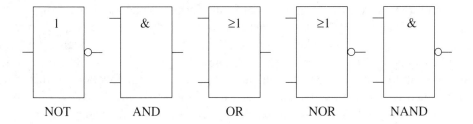

These symbols have found application, but we shall continue to use the American Military Standard symbols, which are the symbols used for logic gates throughout this book. The bubble signifying inversion can actually be attached either to inputs or to outputs, giving two possible representations for an inverter:

NOT   alternative NOT symbol

This is part of a drawing convention known as *mixed logic*, a convention for drawing circuit diagrams of which you can find details in Appendix 1.1.

2.4.6   Complete set of 2-input logic functions

For a 2-input logic gate there are four possible input combinations. For each of the combinations there are two possible output values. A 2-input logic gate then has $2^4 = 16$ possible output sets:

Inputs		Possible output sets															
A	B	0	1	2	3	4	5	6	7	8	9	10	11	12	13	14	15
0	0	0	0	0	0	0	0	0	0	1	1	1	1	1	1	1	1
0	1	0	0	0	0	1	1	1	1	0	0	0	0	1	1	1	1
1	0	0	0	1	1	0	0	1	1	0	0	1	1	0	0	1	1
1	1	0	1	0	1	0	1	0	1	0	1	0	1	0	1	0	1

By inspection we can recognise some of these. Set 1 represents AND, set 7 is the OR function, set 8 is NOR and set 14 is NAND. Some are evidently recognisable and equally evidently not logic gates; set 0 is '0' for all combinations, while set 15 is '1' for all combinations. As logic functions they are called FALSE and TRUE respectively; there is no gate that implements them. Set 3 is a direct copy of A, set 5 is B, set 10 is $\overline{B}$ and set 12 is $\overline{A}$. The existence of these as a function is a consequence of determining all possible output combinations. Others are equally simple: set 2 is '1' in one case, for A = '1' and B = '0'. This is then $A \cdot \overline{B}$ and so set 2 represents an AND gate with one input inverted. So does set 4, $\overline{A} \cdot B$. Set 11 is $A + \overline{B}$ and set 13 is $\overline{A} + B$; again there is no gate to implement them. Only two combinations remain, set 6 and set 9, and for combinational logic design these are functions in their own right. Set 6 represents OR but excluding the case when the inputs are both '1'; in consequence it is called *exclusive OR* or *EXOR*. Set 9 is the logical inversion of set 6 and is called *exclusive NOR* or *EXNOR*. The EXNOR function is '1' when the inputs are the same and for this reason is often called *equivalence*. All functions in this table are actually named in discrete mathematics, but except for AND, NAND, OR, NOR or NOT, only EXNOR and EXOR have any consequence in combinational logic design.

EXOR   Truth table

A	B	$f = A \oplus B$
0	0	0
0	1	1
1	0	1
1	1	0

Symbol

EXNOR   Truth table

A	B	$f = \overline{A \oplus B}$
0	0	1
0	1	0
1	0	0
1	1	1

Symbol

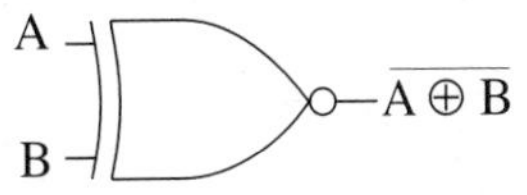

Note that

$$A \oplus B = \overline{A} \cdot B + A \cdot \overline{B}, \overline{A \oplus B} = \overline{A} \cdot \overline{B} + A \cdot B$$

Buffer

The final member of the set of logic functions is a *buffer*, a circuit that does not change the input:

Truth table

Input	Output
0	0
1	1

symbol

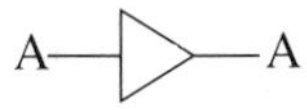

The function clearly has no logic use. It is used to distribute and *interface* signals in logic circuits. Its symbol is that of an amplifier, and that is its function. Logic gates can only source limited amounts of current, and thus only drive a limited number of gates. If we need to distribute a signal via a particular gate, to more than its specified maximum, then we need to buffer the signal and feed it via the buffer circuits. This coincidentally explains the use of the bubble to signify inversion – by its omission, the buffer is not NOT! These complete the full set of available gates.

2.5 Integrated circuits

The AND, OR and NOT gates are available in *integrated circuits* or *chips* which give you packages of gates. The circuit is actually inside the chip and the legs provide connections to your circuit:

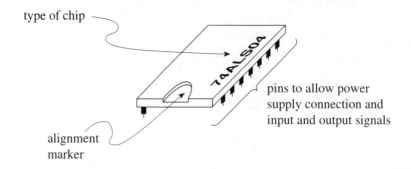

type of chip

pins to allow power
supply connection and
input and output signals

alignment
marker

There is an alignment marker to ensure that you use a chip the right way round. There are two pins for the *power supply* and the remainder are for connections to the circuits inside. There are many varieties of chips; common circuits are usually available in a *logic family* which is a series of common logic functions implemented using a particular *logic technology*. The most famous and enduring logic family is the 74 series introduced by Texas Instruments. Among its members is the 7404, which provides six NOT gates and is commonly known as a hex inverter. The six NOT gates together with two power supplies result in a 14-pin chip, arranged as:

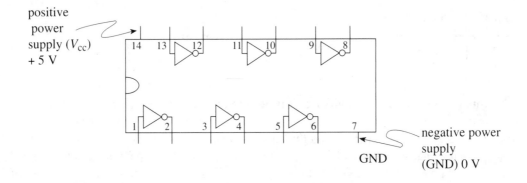

positive
power
supply (V_{cc})
+5 V

negative power
supply
(GND) 0 V

GND

The 74 series is particularly enduring in that it has been implemented in differing logic technologies (TTL and CMOS). Any logic technology is a compromise between many factors, of which speed and power are among the most important.

The 7400 provides four NAND gates (quad NAND) arranged as follows:

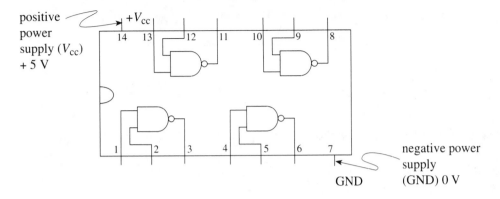

The 74 series includes many more chips, which extend to combinations of gates. Less use is made of these chips now, as *programmable logic* now dominates implementation. Programmable logic chips offer a set of gates and the interconnection between these gates can be specified by software, thus controlling the chip's function. The chip then has a clearly documented design, and reprogrammable chips can easily be reconfigured.

2.6 Basic logic devices

2.6.1 Coders and decoders

The function of a decoder is naturally to decipher encoded information. Information is encoded primarily to maximise usage of channel capacity, such as for satellite communication where the available power is limited. By way of example, consider a traffic light control system where a controller determines the sequence and timing of the lights, but then needs to turn them on and off.

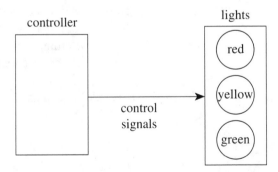

The controller could provide three signals, one to turn on each light. To save money we could use just two wires and transmit logic signals by noting that two logic signals have four possible combinations. We can then design a *coder* which encodes the chosen information. We can choose the signals as

| | | Transmitted signals | |
		A	B
Illuminate red bulb	R	0	0
Illuminate yellow bulb	Y	0	1
Illuminate green bulb	G	1	0

In software this would be of the form

```
IF red     THEN BEGIN
                A=LOW
                B=LOW
            END
IF yellow  THEN BEGIN
                A=LOW
                B=HIGH
            END
IF green   THEN BEGIN
                A=HIGH
                B=LOW
            END
```

We can then use two lines to communicate between the controller and the lights. The signals need to be decoded from the chosen coding scheme to light the correct bulb. We then need a *decoder:*

| Received signals | | Illuminated light | |
A	B		
0	0	R	illuminate red bulb
0	1	Y	illuminate yellow bulb
1	0	G	illuminate green bulb
1	1	X	use don't-care for the unused state

This uses the same format as the encoder to give a decoder circuit as

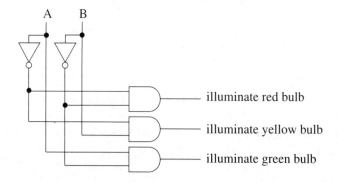

2.6.2 2–4 line decoder

A decoder that decodes the four possible combinations of two input signals is known as a 2–4 line decoder. Its function is given by

Decoder circuit

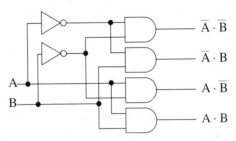

Input signals A B	Output signal channel
0 0	0
0 1	1
1 0	2
1 1	3

A software description of the 2–4 line decoder is

```
CASE (A,B)
BEGIN
     00: BEGIN channel_0=1 ;If A=0 AND B=0 THEN channel_0=1
              channel_1=0
              channel_2=0 ;The others are 0
              channel_3=0 END
     01: BEGIN channel_1=1 ;If A=0 AND B=1 THEN channel_1=1
              channel_0=0
              channel_2=0 ;The others are 0
              channel_3=0 END
     10: BEGIN channel_2=1 ;If A=1 AND B=0 THEN channel_2=1
              channel_0=0
              channel_1=0 ;The others are 0
              channel_3=0 END
     11: BEGIN channel_3=1 ;If A=1 AND B=1 THEN channel_3=1
              channel_0=0
              channel_1=0 ;The others are 0
              channel_2=0 END
END
```

2.6.3 Multiplexers and demultiplexers

The function of a *multiplexer* is to choose an output from a selection of inputs in a manner similar to a rotary switch, which for two inputs is

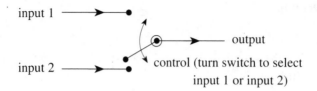

A multiplexer is symbolised by

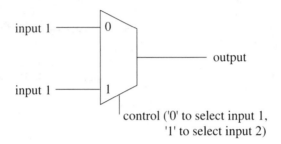

Its function is described as

```
IF (control) THEN output=input_2
  ;control = 1 connects input_2 to the output
            ELSE output=input_1
  ;control = 0 connects input_1 to the output
```

The function of a *demultiplexer* is the reverse, to choose an output signal

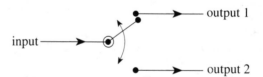

and is symbolised by

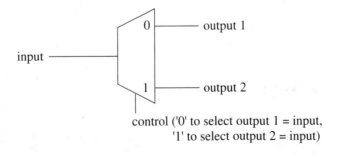

In software its function is

```
IF (control) THEN output_2=input
   ;control = 1 connects output_2 to the input
            ELSE output_1=input
   ;control = 0 connects output_1 to the input
```

There are other forms of multiplexing, such as time domain multiplexing where signals occupy a communication channel for specified time slots (e.g. in radio communications). The function of the multiplexer is then to choose from the input signals at the appropriate time slot.

2.6.4   4-1 line multiplexer

A circuit that chooses a single output from one of four inputs is called a 4–1 line multiplexer. Note that we need to choose which of four lines is to be selected and then provided as the circuit output. Two logic signals provide four combinations and the channel to be chosen is then encoded using two select lines. These lines are given by

Select lines		Channel selected
A	B	
0	0	0
0	1	1
1	0	2
1	1	3

A software description is

```
CASE (A,B)
BEGIN
   00: output=channel_0 ;If A=0 AND B=0 THEN output=channel_0
   01: output=channel_1 ;If A=0 AND B=1 THEN output=channel_1
   10: output=channel_2 ;If A=1 AND B=0 THEN output=channel_2
   11: output=channel_3 ;If A=1 AND B=1 THEN output=channel_3
END
```

The 4–1 line multiplexer is symbolised by

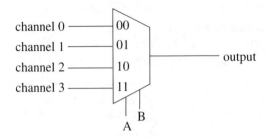

Note that the specification on channel selection is identical to that for the 1–4 line decoder. This is hardly surprising, because we are decoding two input lines to determine which of four inputs should be selected as output. The decoder signals are enabled when a particular channel is chosen and the multiplexer circuit is then given by ANDing the decoder outputs with the appropriate input signal. The outputs of the AND gates are then combined with a 4-input OR gate whose output will follow any selected channel:

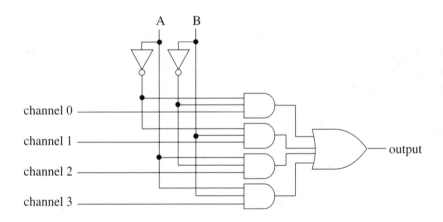

The function is then defined by a truth table:

Input channel status				Select lines		Output	Channel
0	1	2	3	A	B		
0	X	X	X	0	0	0	} 0
1	X	X	X	0	0	1	
X	0	X	X	0	1	0	} 1
X	1	X	X	0	1	1	
X	X	0	X	1	0	0	} 2
X	X	1	X	1	0	1	
X	X	X	0	1	1	0	} 3
X	X	X	1	1	1	1	

2.7 Introductory combinational design

The design of a circuit can be extracted from a truth table specification. This is to be expected, since a truth table is a complete specification of a circuit's operation. In a basic form the design can be implemented using AND, OR and NOT gates.

The basis for such design is that the function will be '1' for various combinations of inputs. These input combinations can be grouped using AND gates, since the output of an AND gate is '1' when all its inputs are a '1'. The output of the circuit should be '1' if any of these combinations produces a '1', and we can therefore group the AND gate outputs using an OR gate, since the output of an OR gate is '1' if any of its inputs is '1'. This procedure is reflected in the earlier circuit for the multiplexer where the gated decoder outputs were collected in an OR gate. The procedure for design is then

1. Fill in the truth table.
2. Extract conditions for which the output = '1'.
3. Group the conditions to form the final output.

The process is best illustrated by example. Consider the following 'specification'.

> John Wonderland is going on holiday with his wife Alice, his beautiful daughter Bo and her friends, Chas and Dave. He wants to fish peacefully while his wife chaperones Bo, so he decides to make an alarm which goes off when Bo is with either Chas or Dave (or both) but not Alice, either inside or outside the house. He buys radio transmitters to indicate whether or not people are inside the house and surreptitiously fixes them to everyone. He then needs an alarm circuit to act on the information from the transmitted signals.

This is clearly a fictitious problem. My students have suggested that they would steal Alice's alarm and fix it to Bo.

We shall follow the stated design procedure; the first step is to draw up the truth table.

We shall use   A = Alice, B = Bo, C = Chas, D = Dave and f = alarm;
and for all inputs   '1' = inside house, '0' = outside house;
and for the alarm   '1' = alarm rings, '0' = alarm is silent.

The truth table is then

A	B	C	D	f		
0	0	0	0	0		all outside together
0	0	0	1	0		Alice and Bo outside together
0	0	1	0	0		Alice and Bo together
0	0	1	1	0		Alice and Bo together
0	1	0	0	0		Bo inside alone
0	1	0	1	1	f_1	Bo inside with Dave
0	1	1	0	1	f_2	Bo inside with Chas
0	1	1	1	1	f_3	Bo with Chas and Dave
1	0	0	0	1	f_4	Bo outside with Chas and Dave
1	0	0	1	1	f_5	Bo outside with Chas only
1	0	1	0	1	f_6	Alice inside, Bo outside with Dave
1	0	1	1	0		Bo outside alone
1	1	0	0	0		Bo and Alice inside
1	1	0	1	0		Alice, Bo and Dave inside
1	1	1	0	0		Alice, Bo and Chas inside
1	1	1	1	0		all inside

The output is true in six of the sixteen cases; in all others Bo is chaperoned by Alice, or alone, and the alarm does not go off. These six cases are labelled f_1, f_2, f_3, f_4, f_5 and f_6. In each of these cases a combination of inputs forces the output to be '1'.

For f_1 the output = '1' for A = '0', B = '1', C = '0', and D = '1'
i.e. when $\quad\quad\quad\quad \overline{A}$ = '1', B = '1', $\overline{C}$ = '1', and D = '1'

f_1 is now expressed as variables which are all '1', which can thus be grouped using a 4-input AND gate. f_1 can now be implemented as

$$f_1 = \overline{A} \cdot B \cdot \overline{C} \cdot D$$

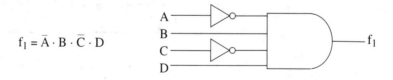

and the output will be '1' when Alice and Chas are outside and Bo and Dave are inside. Note that f_1 will be '0' for all other input combinations.

We can do the same for f_2:

f_2 = '1' for $\quad\quad$ A = '0', B = '1', C = '1' and D = '0'
i.e. when $\quad\quad \overline{A}$ = '1', B = '1', C = '1' and $\overline{D}$ = '1'

which can be implemented as

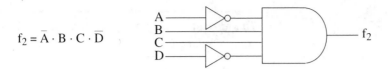

$$f_2 = \overline{A} \cdot B \cdot C \cdot \overline{D}$$

which is '1' only when Alice and Dave are outside, and Bo is inside with Chas.
  Similarly

$$f_3 = \overline{A} \cdot B \cdot C \cdot D, \quad f_4 = A \cdot \overline{B} \cdot \overline{C} \cdot \overline{D}, \quad f_5 = A \cdot \overline{B} \cdot \overline{C} \cdot D, \quad \text{and} \quad f_6 = A \cdot \overline{B} \cdot C \cdot \overline{D}$$

We now have six circuits which provide the alarm if their particular input combination is satisfied. We want to collect together the six terms to provide a single alarm. We need a circuit whose output is '1' when any of the inputs are a '1' – this is an OR gate. We can then collect f_1, f_2, f_3, f_4, f_5 and f_6 together, using

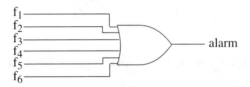

This then gives the full alarm circuit to provide the specified alarm according to the location of the conspirators.

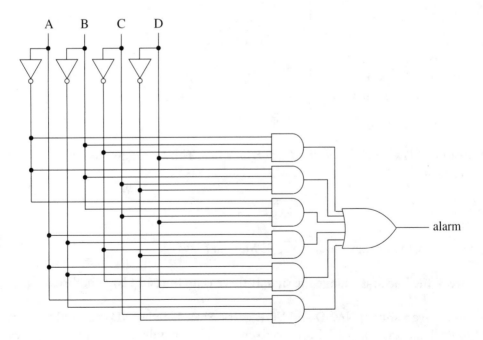

> *This circuit uses more gates than it needs, and it can be reduced or minimised (see Section 2.9).*

A software description for the alarm function for the same inputs is

```
IF    ((NOT(alice) AND bo AND chas AND dave)
   OR (NOT(alice) AND bo AND chas AND NOT(dave))
   OR (NOT(alice) AND bo AND NOT(chas) AND dave)
   OR (alice AND NOT(bo) AND NOT(chas) AND NOT(dave))
   OR (alice AND NOT(bo) AND NOT(chas) AND dave)
   OR (alice AND NOT(bo) AND chas AND NOT(dave)))
       THEN alarm=TRUE
       ELSE alarm=FALSE
```

2.8 Laws of logic

We have so far defined logic without a formal algebra. Including the algebra is perhaps pedantic, but much logic use does depend on it. Given logical inversion, '$\overline{0}$' = '1', $\overline{\overline{A}} = A$

OR laws	AND laws
$A + 0 = A$	$A \cdot 0 = 0$
$A + 1 = 1$	$A \cdot 1 = A$
$A + \overline{A} = 1$	$A \cdot \overline{A} = 0$
$A + B = B + A$	$A \cdot B = B \cdot A$
$A + (B + C) = (A + B) + C$	$A \cdot (B \cdot C) = (A \cdot B) \cdot C$
$A + B \cdot C = (A + B) \cdot (A + C)$	$A \cdot (B + C) = A \cdot B + A \cdot C$
$\overline{A + B} = \overline{A} \cdot \overline{B}$	$\overline{A \cdot B} = \overline{A} + \overline{B}$

The last two laws are versions of *de Morgan's law*. This law shows us how to swap logic implementation.

$$\overline{\overline{A} + \overline{B}} = A \cdot B \qquad\qquad \overline{\overline{A} \cdot \overline{B}} = A + B$$

An easy way to remember how to use de Morgan's law is:

Break the line and change the sign. If there is no line, add two and break one.

It implies **universality of NAND and NOR gates**. Since both NAND and NOR gates can be used to implement OR and AND, respectively, by inverting the inputs any function can be implemented using NAND or NOR gates only. This is illustrated further in Section 2.11.

De Morgan's law is true semantically, too. Compare 'I cannot water the garden with a leaky bucket and no water' with 'I can only water the garden with a good bucket and water'. You can therefore use de Morgan's law to simplify complicated IF statements when programming.

Two other laws can be derived from this algebra:

$$A + A \cdot B = A \qquad A + \overline{A} \cdot B = A + B$$

For proof, for the first of these

$$A + A \cdot B = A \cdot (1 + B) = A \cdot 1 = A$$

for the second, by substitution for $A + A \cdot B = A$, then

$$A + \overline{A} \cdot B = A + A \cdot B + \overline{A} \cdot B = A + B(A + \overline{A}) = A + B \cdot 1 = A + B$$

These give reduced circuits:

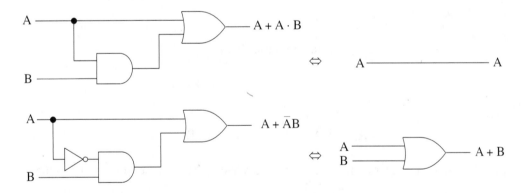

In other words, why use gates when they are redundant? Each gate costs money and has a certain reliability and lifetime associated with it. If we can reduce the number of gates we save money and effort. This is *minimisation*.

Formal proof is more involved and rigorous. This is a functional proof only.

De Morgan's law can also be used to explain two symbols in common use for NOR and NAND. These are due to the fact that NAND is equivalent to the OR function with inverted inputs and NOR is equivalent to AND with inverted inputs. By de Morgan's law, and by using a circle on the input connection to denote logical inversion (as in the mixed logic convention) we obtain:

$$\overline{A \cdot B} = \overline{A} + \overline{B} \text{ (NAND becomes OR)} \quad \text{and} \quad \overline{A + B} = \overline{A} \cdot \overline{B} \text{ (NOR becomes AND)}$$

| NAND | Equivalent NAND symbol | NOR | Equivalent NOR symbol |

2.9 Minimisation

Circuits cost money and consume power (also costing money). If we reduce the circuit, we reduce the cost (increase profit). Minimisation is therefore prompted by the desire to implement a logic function in as few gates as possible to reduce cost. Designers now use logic minimisers (implemented via computer programs) which do this automatically. We shall study some by-hand techniques. These are mainly historical now, but necessary because logic minimisers develop from them. There are *algebraic* and *graphical* techniques.

2.9.1 Algebraic minimisation

Earlier rules give the basis for algebraic minimisation:

$$A + \overline{A} = 1 \quad \text{so} \quad B \cdot (A + \overline{A}) = B$$

Note:

$$A + A \cdot B = A \quad \text{and} \quad A + \overline{A} \cdot B = A + B$$

John Wonderland's alarm circuit can now be reduced. The final extraction from the truth table was

$$f = \overline{A} \cdot B \cdot \overline{C} \cdot D + \overline{A} \cdot B \cdot C \cdot \overline{D} + \overline{A} \cdot B \cdot C \cdot D + A \cdot \overline{B} \cdot \overline{C} \cdot \overline{D} + A \cdot \overline{B} \cdot \overline{C} \cdot D + A \cdot \overline{B} \cdot C \cdot \overline{D}$$

$$\overline{A} \cdot B \cdot D \cdot (\overline{C} + C) \quad \overline{A} \cdot B \cdot C \cdot (\overline{D} + D) \quad A \cdot \overline{B} \cdot \overline{D} \cdot (\overline{C} + C) \quad A \cdot \overline{B} \cdot \overline{C} \cdot (\overline{D} + D)$$

which using common terms gives

$$f = \overline{A} \cdot B \cdot D + \overline{A} \cdot B \cdot C + A \cdot \overline{B} \cdot \overline{D} + A \cdot \overline{B} \cdot \overline{C}$$

which can be implemented using 2-input gates by sharing common terms

$$f = \overline{A} \cdot B \cdot (C + D) + A \cdot \overline{B} \cdot (\overline{C} + \overline{D})$$

and the minimised circuits are

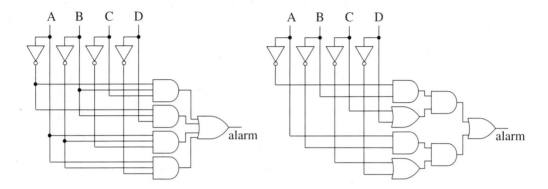

We have then taken a function which required four inverters, six 4-input gates and one 6-input OR gate to seven 2-input gates plus the inverters. Note that you can consider terms more than once for the purposes of minimisation, because $A + A = A$, i.e. adding the term back in does not change the function at all. Algebraic techniques are limited to small-scale problems; for circuits with more than five inputs they become tedious and error-prone.

2.9.2 *Graphical minimisation – the Karnaugh map*

The Karnaugh map (K-map) uses $A + \overline{A} = A$ in a graphical format. Essentially, it is a 2-dimensional truth table. We first draw a box with inputs around it:

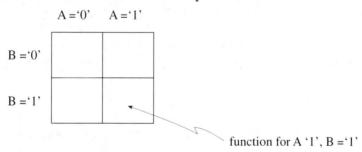

Then we insert the function as for each input combination; a K-map for AND is

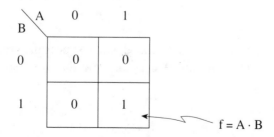

This shows that the K-map is an alternative formulation of the truth table; it is now expressed in a compact two-dimensional way. But it has been introduced as a graphical minimisation technique, so how do we minimise using this? Consider the K-map for the OR function:

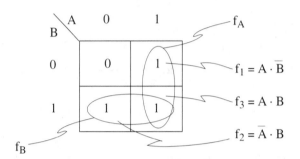

By using algebraic minimisation based on f_1, f_2 and f_3 (including f_3 twice) we obtain

$$f = A \cdot \overline{B} + \overline{A} \cdot B + A \cdot B = A \cdot (B + \overline{B}) + (\overline{A} + A) \cdot B = A + B$$

Extraction from the truth table was based on using individual terms and then reducing them algebraically. The K-map enables this reduction to be spotted **visually** – we can see terms of the form $A \cdot B + A \cdot \overline{B} = A \cdot (B + \overline{B}) = A \cdot 1 = A$. We can draw loops around these common terms to minimise the circuit. From the K-map for OR the function is then described by two loops, f_A and f_B, and since $f_A = A \cdot (B + \overline{B}) = A$ and $f_B = B \cdot (A + \overline{A})$, then the function is the sum of the two loops. The function is then $f = f_A + f_B$, so $f = A + B$ as expected. A group of two '1's removes a single variable by covering both possible states of that variable, as in

$$A \cdot (B + \overline{B}) = A \cdot 1 = A$$

A group of four removes two variables by covering all four possible combinations as

$$A \cdot (\overline{B} \cdot \overline{C} + \overline{B} \cdot C + B \cdot \overline{C} + B \cdot C) = A \cdot (\overline{B} \cdot (\overline{C} + C) + B \cdot (\overline{C} + C))$$
$$= A \cdot (\overline{B} + B) = A$$

To achieve the minimised circuit you should take the largest possible groups (of 2, 4, 8) to reduce the number of inputs in each extracted term. Note that the extracted groups can overlap.

A K-map for three inputs is given by   and for four inputs

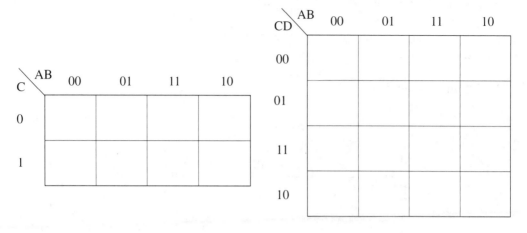

Note the order of the inputs is **not** the same as for a truth table, and the last two columns (or rows) flip from the values expressed in the truth table. This then forces the order of the truth table to reflect the *unit distance criterion* where one input only changes between adjacent cells. This allows minimisation by looping elements to perform

$$A \cdot (\overline{B} + B) = A$$

If we were to order inputs as in a truth table then this grouping would not perform minimisation.

Note also that the K-map wraps round the sides and around the top. The left column is for inputs $AB = 00$ and the rightmost column for $AB = 10$. These then still obey the unit distance criterion and if two elements in them were a '1' then they can be looped around the box to reduce and remove the input A. The map also wraps round from top to bottom since only the input C changes between the top and bottom rows. Note that the map is not a sphere; it wraps round only from top to bottom, and from side to side. Some find a K-map easier when it is drawn as

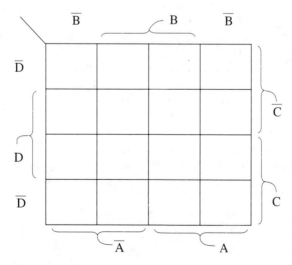

This is identical to the map drawn before.

2.9.3   *Examples of Karnaugh map design*

Use of the K-map is illustrated by examples showing non-minimised and minimised solutions.

Minimised solution?   Final extraction

(1)

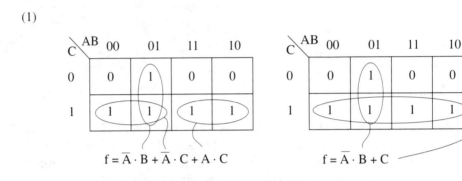

$$f = \overline{A} \cdot B + \overline{A} \cdot C + A \cdot C \qquad f = \overline{A} \cdot B + C$$

Here, a larger group should have been taken in the first extraction which is reflected in the possibility of algebraically minimising the K-map result. This should not occur since the K-map is designed to avoid it. The second extraction results in a minimised solution.

(2)

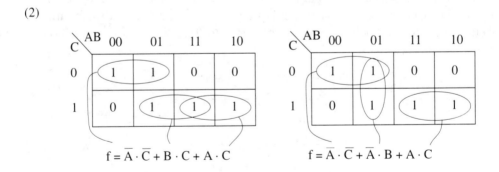

$$f = \overline{A} \cdot \overline{C} + B \cdot C + A \cdot C \qquad f = \overline{A} \cdot \overline{C} + \overline{A} \cdot B + A \cdot C$$

In the second example, both extractions implement the circuit in minimised form. There is no unique solution.

(3)

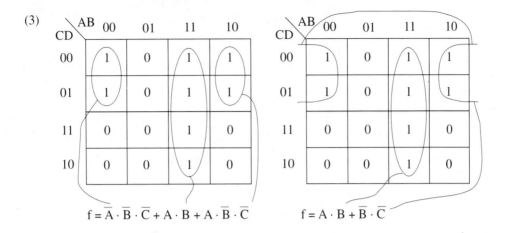

$$f = \overline{A} \cdot \overline{B} \cdot \overline{C} + A \cdot B + A \cdot \overline{B} \cdot \overline{C}$$

$$f = A \cdot B + \overline{B} \cdot \overline{C}$$

In the third extraction, we have not remembered that the table maps round from side to side and a better, more minimised extraction is possible.

(4)

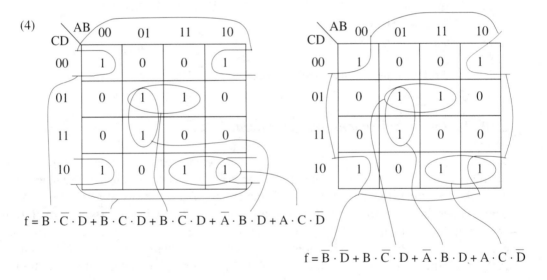

$$f = \overline{B} \cdot \overline{C} \cdot \overline{D} + \overline{B} \cdot C \cdot \overline{D} + B \cdot \overline{C} \cdot D + \overline{A} \cdot B \cdot D + A \cdot C \cdot \overline{D}$$

$$f = \overline{B} \cdot \overline{D} + B \cdot \overline{C} \cdot D + \overline{A} \cdot B \cdot D + A \cdot C \cdot \overline{D}$$

Finally, in the fourth example, in its non-minimised form we have remembered that the map wraps round from side to side, but not from top to bottom. The four corner elements therefore form a group which can be extracted together as $\overline{B} \cdot \overline{D}$. Finally, consider the following K-map:

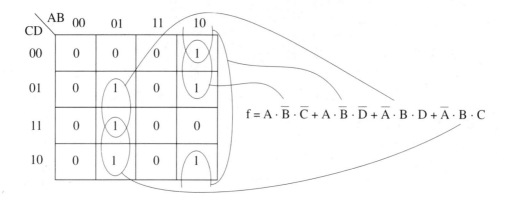

This is John Wonderland's alarm circuit again. Now the minimisation is clear and easy. A minimised software description for John's alarm is:

```
IF   ((NOT(alice) AND bo AND (chas OR dave))
  OR   (alice AND NOT(bo) AND (NOT(chas) OR NOT(dave)))
       THEN alarm=TRUE
       ELSE alarm=FALSE
```

This clearly follows the original specification.

2.10 Timing considerations and static hazards

Any circuit takes time to respond. For gates the *propagation delay* defines how long before the output changes state after an input. The time for the output to change LOW to HIGH, T_{PLH}, or the time for HIGH to LOW transition, T_{PHL}, are usually of the order of nanoseconds. It is therefore important to consider how many logic gates signals pass through (how much delay there is in the circuit). This is called the *level of the logic system* (not to be confused with logic levels, which are the ranges of acceptable voltages for a '1' and a '0'). A 2-level logic system is one where a signal will pass through two gates at maximum from the input to the output. This can be important in considering the **delay** introduced by combinational logic via T_{PHL} or T_{PLH}.

A *hazard* is clearly undesirable in a logic system. It concerns a logic variable at an **unexpected** state that is often due to **timing considerations** (propagation delay). The propagation delay of an inverter can cause a momentary *glitch* on a circuit output. Consider

$$f = A \cdot \overline{B} + B \cdot C$$

We shall look at the output state for inputs A = '1' and C = '1' and the input B changes from '1' to '0'. The circuit output is the OR of two terms, either of which is '1' when B changes, and so the output should not change at all.

This is true in theory, but in practice $\overline{B}$ is derived from an inverter whose input is B and when B changes from '1' to '0' then the output of the inverter, $\overline{B}$, changes from '0' to '1' T_{PLH} ns later. There is then a short time during which both $A \cdot \overline{B}$ and $B \cdot C$ are zero. The output will then be momentarily '0', while the change in B propagates through the inverter. This is unwelcome but it is reality, since no circuit can respond infinitely fast; it is called a *static hazard*.

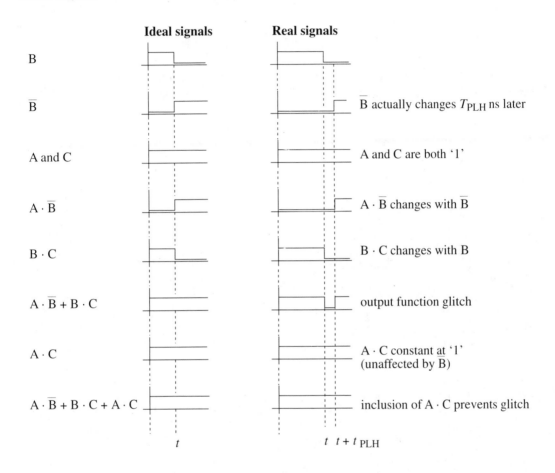

To reduce its effect we can link together all the largest groups extracted for minimisation. The extra terms we introduce are called *bridging terms*. The K-map for

$$f = A \cdot \overline{B} + B \cdot C$$

shows how the bridging term is derived:

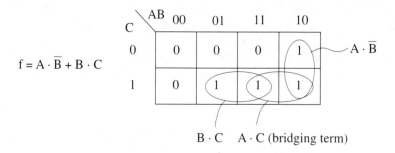

$$f = A \cdot \overline{B} + B \cdot C$$

The term $A \cdot C$ is not affected by the change in B, and its inclusion in the overall function removes the glitch due to the static hazard. The hazard-free implementation is then

$$f = A \cdot \overline{B} + B \cdot C + A \cdot C$$

Further examples are:

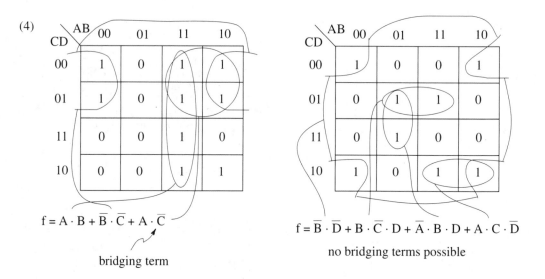

In practice the treatment of hazards is more complex. There are some hazards that are not exposed by determining the bridging terms in K-maps, and it is often not possible to link all terms together. It is often better to **wait** until any timing components should be settled, i.e. expect them as part of life and design around them. The time for which you should wait is given by the worst-case path through the logic, i.e. the path that a signal takes the longest time to traverse.

2.11 Logic implementations

De Morgan's law implies that NAND and NOR gates are universal. This can be illustrated via a K-map where extracting the '1's gives an AND/OR function, and by extracting the '0's we achieve AND/NOR, e.g.:

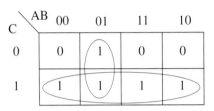

By extracting the '1's we achieve

$$f = \overline{A} \cdot B + C$$

This can be implemented using NAND gates only by application of de Morgan's law, thereby illustrating the universality of NAND gates.

$$f = \overline{\overline{\overline{A} \cdot B} \cdot \overline{C}}$$

and by extracting the '0's we achieve

$$\overline{f} = A \cdot \overline{C} + \overline{B} \cdot \overline{C}$$

i.e.

$$f = \overline{A \cdot \overline{C} + \overline{B} \cdot \overline{C}}$$

is an AND/NOR function. We can again achieve an entirely NOR construction by applying de Morgan's law, again re-illustrating that NAND and NOR gates are both universal:

$$f = \overline{\overline{\overline{A} + C} + \overline{\overline{B} + C}}$$

By applying de Morgan's law again,

$$f = (\overline{A} + C) \cdot (B + C)$$

$$= \overline{A} \cdot B + B \cdot C + \overline{A} \cdot C + C$$

$$= \overline{A} \cdot B + C$$

This returns us to the function extracted by the '1's, hence showing the duality of the two extractions.

Using NAND or NOR to provide NOT

Since gates can come in packages it is sometimes convenient to deploy a NAND gate or a NOR gate as a NOT gate to save using an extra chip. From the truth tables for NAND and NOR,

A	B	$\overline{A \cdot B}$	$\overline{A+B}$
0	0	1	1
0	1	1	0
1	0	1	0
1	1	0	0

By inspection we can see that $\overline{A \cdot B} = \overline{A}$ if the other input B equals A (i.e. the two inputs are connected together) or if B = '1'. Similarly $\overline{A + B} = \overline{A}$ if B = '0' or B = A, giving four possible circuits equivalent to NOT:

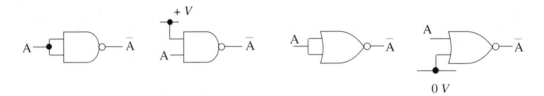

The circuits where one input is connected low or high are usually preferred, since the input is connected to a power supply, not to the output of another gate. The connection to the power supply is usually made using a resistor for practical reasons.

2.12 Terminology

The extraction of a function by grouping inputs using AND gates whose outputs are grouped using OR is called the *sum of products* (SOP) form, e.g.

$$f = \overline{A} \cdot \overline{B} \cdot \overline{C} + A \cdot B \cdot \overline{C}$$

It conforms rather nicely with the way we think about things. There is conversely a *product of sums*, e.g.

$$f = (\overline{A} + B) \cdot (A + C)$$

A product term containing one occurrence of **every** variable is known as a *minterm* or a *canonical product term*. A sum term containing one occurrence of **every** variable is known as a *maxterm* or a *canonical sum term*. A function expressed as OR of distinct minterms is known as a *canonical sum of products*; one expressed as AND of distinct maxterms is known as a *canonical product of sums*. For those with masochistic tendencies, a canonical SOP is a *disjunctive normal form*, while a canonical POS is a *conjunctive normal form!* For further developments let us consider a function given by a truth table as follows:

Decimal	J	K	L	f
0	0	0	0	0
1	0	0	1	0
2	0	1	0	0
3	0	1	1	1
4	1	0	0	1
5	1	0	1	1
6	1	1	0	1
7	1	1	1	1

where the decimal number corresponds to the binary representation of the input coding. This function can be expressed using a decimal index for each term in the truth table. The minterms are the cases for which the function is '1', and collecting these in sum of products (SOP) form using the decimal index is

$$f = \bar{J} \cdot K \cdot L + J \cdot \bar{K} \cdot \bar{L} + J \cdot \bar{K} \cdot L + J \cdot K \cdot \bar{L} + J \cdot K \cdot L$$

which can be expressed in a shortened manner as

$$f = \sum_{J,K,L}(3,4,5,6,7)$$

It can also be expressed as a product of sums

$$\bar{f} = \bar{J} \cdot \bar{K} \cdot \bar{L} + \bar{J} \cdot \bar{K} \cdot L + \bar{J} \cdot K \cdot \bar{L}$$

hence

$$f = \overline{\bar{J} \cdot \bar{K} \cdot \bar{L} + \bar{J} \cdot \bar{K} \cdot L + \bar{J} \cdot K \cdot \bar{L}}$$

Now, by de Morgan's law

$$f = \overline{\bar{J} \cdot \bar{K} \cdot \bar{L}} + \overline{\bar{J} \cdot \bar{K} \cdot L} + \overline{\bar{J} \cdot K \cdot \bar{L}}$$

and again by de Morgan's law

$$f = (J+K+L) \cdot (J+K+\bar{L}) \cdot (J+\bar{K}+L)$$

$$= \prod_{JKL}(0,1,2)$$

This is the canonical product of sums specification. Note the 'inversion' in canonical POS expression and the relation between SOP and POS. Logic functions are often expressed in POS and SOP form – it is a very compact description, and one not prone to misinterpretation. By K-map for the original function f,

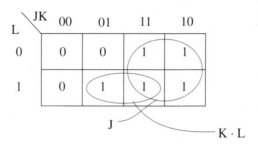

so $f = J + K \cdot L$ – this is the condensed or minimised form (i.e. a *minimal sum*). The terms J and $K \cdot L$ are non-canonical (they do not contain all the variables) and are called *prime implicants*; these are the largest loops you can draw in the K-map. For a K-map containing four prime implicants,

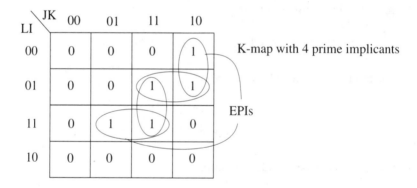

K-map with 4 prime implicants

EPIs

Note that the minimal sum does not necessarily contain all prime implicants; a hazard-free one does. Terms which must be included in the minimal sum are called *essential prime implicants* (EPIs on the diagram) and are the only prime implicants that cover a particular '1' cell. Other '1's may be covered by two or more prime implicants, **one** of which must be included in the minimal sum.

2.13 Concluding comments and further reading

In this chapter we have seen the development of logic from its formal status to a circuit implementation. Many of the themes presented in this chapter will be manifest later in this book. One major consideration is the use of software to describe circuit operation. This is now widely used in logic, and we shall be using it later in **synthesis**, i.e. to build logic circuits. The development of minimisation has studied some by-hand techniques. The algebraic techniques actually develop into computer programs which can be used for circuits with large numbers of inputs. The K-map is very restricted in this respect. Imagine drawing a K-map for a function of ten inputs with 1024 spaces; then you start finding the loops! It is sufficient as a hand tool and gives worthwhile insight, but there its use ends. Finally, we have seen some of the basic building blocks and have been introduced to the design of combinational logic systems. These will be used throughout the book, particularly in the design examples.

Two books in particular give the historical basis of logic: Boole, G., *An Investigation into the Laws of Thought*, originally published in 1854, and republished in 1954 by Dover Publications, New York, and Carroll, L., *Game of Logic*, originally published in 1896, and republished, again by Dover Publications, in 1958. These might be worth a look if you can find copies. The discrete mathematics approach to logic is ably expounded in Wiitala, S.A., *Discrete Mathematics: A Unified Approach* (McGraw-Hill, 1987).

There is a plethora of texts on digital electronics, though none employs a software approach. Texts in this area can date very rapidly. Among those which have found popularity are: Wakerly, J. F., *Digital Design: Principles and Practices* (Prentice-Hall, 2nd edn, 1994) and Unger S. H., *The Essence of Logic Circuits* (Prentice-Hall, 1989). These are weighty texts which cover a great amount of material, but perhaps in rather excessive detail.

Minimisation is a topic of continuing interest and research. The Quine–McKlusky algorithm forms the basis of some logic minimisers and warrants further study, though not in this introductory text (see for example Brayton, R.K., Hachtel, G.D., McMullen, C. T. and Sangiovanni-Vincentelli, A. L., *Logic Minimisation Algorithms for VLSI Synthesis* (Kluwer Academic Publishers, 1985)). Another topic of major interest is the specification of circuits and their operation. A software specification clearly expresses this, which is itself another avenue for further investigation.

Those who would like a more detailed approach to programming should try Deitel, H. M. and Deitel, P. J., *How to Program* (Prentice-Hall, 1992) which for those who would like more background on electronic circuit theory should read Hayt, W. H. and Kemmerly, J. E., *Engineering Circuit Analysis* (McGraw-Hill, 5th edn, 1986).

2.14 Questions

1 Show that

$$(A + B) \cdot \overline{A \cdot B} = A \oplus B$$

$$\overline{A + B} \cdot A \cdot B = \text{'0'}$$

$$\overline{A} \oplus \overline{B} = A \oplus B$$

$$\overline{\overline{A + B} + \overline{A \cdot B}} = A \cdot B$$

2 Implement the function

$$f = \sum_{ABCD}(4, 6, 9, 11)$$

using a minimum number of standard logic gates (AND, OR, NOT, NAND or NOR) which can have any number of inputs. Implement it again using NOR gates only.

3 Implement the function

$$f = \prod_{ABCD}(0, 2, 4, 5, 6, 7, 11, 15)$$

assuming that complemented inputs are available, but without static hazards using a minimum number of standard logic gates (AND, OR, NOT, NAND or NOR) which can have any number of inputs.

4 Implement the function

$$f = (\overline{B} \cdot \overline{C} \cdot \overline{D}) + (\overline{B} \cdot \overline{C} \cdot \overline{A \cdot D}) + (A \cdot \overline{B} \cdot \overline{C}) + (B \cdot C \cdot (A \oplus D)) + (\overline{A} \cdot B \cdot C) + (B \cdot C \cdot \overline{\overline{A} \cdot \overline{D}})$$

using a minimum number of standard logic gates (AND, OR, NOT, NAND or NOR) and express your solution in sum-of-products form.

5 Implement the function

$$f = (\overline{B} \cdot \overline{D} \cdot \overline{A \oplus C}) + (B \cdot D \cdot \overline{A \oplus C}) + (A \cdot C \cdot \overline{\overline{B} \cdot \overline{D}}) + (\overline{B} \cdot (A \cdot C + \overline{A} \cdot \overline{C})) + (\overline{A} \cdot \overline{C} \cdot B \cdot D)$$

using a minimum number of standard logic gates (AND, OR, NOT, NAND or NOR) and express your solution in product-of-sums form.

6 What is the level of logic associated with the function

$$f = \overline{A \cdot \overline{B} \cdot D + A \cdot C \cdot \overline{D}}$$

when implemented using standard logic gates with any number of inputs? If the T_{PLH} for each combinational logic gate is 10 ns, and T_{PHL} for each combinational logic gate is 11 ns, what is the worst-case propagation delay through the circuit?

7 Implement the function

$$f = \overline{A} \cdot \overline{B} \cdot \overline{C} + \overline{A} \cdot B \cdot \overline{C} + A \cdot B \cdot \overline{C}$$

using an 8-1 line multiplexer.

8 Express the function

$$f = \sum_{ABC}(1, 2, 4, 7)$$

in software with logic functions AND, OR and NOT applied to logic variables that can be assigned HIGH (f = '1') or LOW (f = '0'). Use IF...THEN...ELSE statements only.

9 Express the function in Question 8 using the same logical variable specification, but using CASE statements only.

10 Draw up a truth table for each element of the expression

$$f = \overline{A \cdot C} \cdot (B \oplus D) + A \cdot \overline{C} \cdot (\overline{B \oplus \overline{D}}) + B \cdot \overline{C} \cdot D + \overline{B + C + D}$$

and hence derive a truth table for the function f. Derive a minimised result using a K-map and confirm your result analytically.

11 Design a safety system for a one-seater car, given:

(i) an indicator that provides an output of $+5\,\text{V}$ when someone is sitting in the car and $0\,\text{V}$ otherwise;
(ii) an indicator that provides an output which is $+5\,\text{V}$ when the seat belt is connected properly across the occupant and $0\,\text{V}$ otherwise;
(iii) an indicator showing that the door is shut ($+5\,\text{V}$ output when the door is shut and $0\,\text{V}$ otherwise);
(iv) a system safety condition indicator for the brakes and suspension, with output $+5\,\text{V}$ when the condition is unsafe and $0\,\text{V}$ otherwise.

The safety system should provide a signal f that can be used to allow the car to proceed (f = '1' to allow progress and '0' to prevent progress). The car should not proceed when a driver is not seated in it, or when the driver is not wearing a seat belt, or when the system condition is unsafe. The car should not proceed when the door is open.

(a) Draw up a truth table for the signal f.
(b) Extract the logical function which gives the signal f.
(c) Implement the system using NOT, AND and OR gates.
(d) Implement it again, using NOT and NAND or NOR gates.
(e) Express f in canonical sum-of-products form and in canonical product-of-sums form.

12 A circuit is required which will compare the magnitude of two 2-bit binary numbers, AB and CD. Each number may have a value 0 (when A = '0', B = '0', C = '0' and D = '0'), 1 (when A = '0', B = '1' and C = '0', D = '1'), 2 (when A = '1', B = '0' and C = '1', D = '0'), or 3 (when A = '1', B = '1' and C = '1', D = '1'). The complemented outputs of A, B, C and D are available.

Circuit outputs G, E and L are to be TRUE ('1') when the number AB is greater than, equal to, or less than the number CD, respectively. The outputs should be FALSE ('0') otherwise.

(a) Produce a K-map for each output (G, L and E).
(b) Obtain minimal sum of products expressions for G, L and E.
(c) Implement G using NAND gates only.
(d) Implement E using NOR gates only.
(e) Show how L may be obtained using an 8-1 multiplexer.

13 Identify the static hazards which might be associated with a direct implementation of the following expression

$$f = \overline{A} \cdot C \cdot D + B \cdot C \cdot \overline{D} + A \cdot \overline{C} \cdot D + A \cdot \overline{B} \cdot \overline{C} \cdot \overline{D}$$

Using a K-map show how these static hazards might be removed. The function is required as part of a control system, and the designer needs to know how long after the inputs have changed the output can be used. For all gates T_{PHL} is 9 ns and T_{PLH} is 10 ns. By considering the level of logic, estimate how long after the inputs change, the output of the original function and the output of your hazard-free one can be used by the designer.

3 Logic Circuits

'... the white heat of technological revolution'

attr. Harold Wilson

3.1 Device characteristics

The aim of this chapter is to describe what is inside an integrated circuit. Integrated circuits are made from *switching devices*. These switching devices connect together inputs to implement logic functions. An ideal switching device would respond infinitely fast, consume no power, and be physically small and lightweight. This is practically impossible to achieve, and any switching circuit is a compromise between these factors. The *logic technology* concerns the physical implementation of the switching circuits and offers a distinct set of *performance characteristics*. We shall first investigate the performance characteristics of particular interest in integrated circuit technology, then go on to look at switching circuits, and finally consider the logic technologies that are currently available.

The performance characteristics can be developed from an **ideal** logic gate. The ideal inverter connects its output HIGH to the positive supply rail for a LOW input. For a HIGH input the output is connected LOW to the negative voltage supply, usually ground. The input then acts adversely on a pair of switches (denoted by boxes with logic inputs); one connects the output HIGH, the other connects the output LOW. Thus:

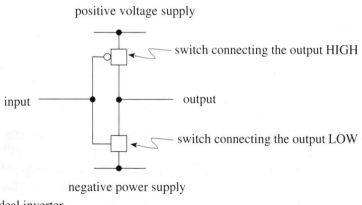

Ideal inverter

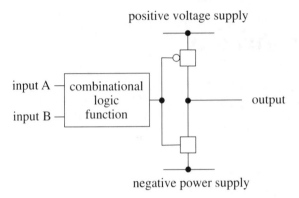

positive voltage supply

input A — combinational logic function

input B —

output

negative power supply

Ideal combinational logic gate

Since the input acts adversely on either switch, the upper switch is turned ON for a LOW input, which is signified by the inversion symbol attached to the input to the switch. Further gates, such as AND and NAND, require appropriate combination of the input signals to control the output switches.

Ideally, the switches connect the output (voltage) level HIGH or LOW in response to an input voltage level. We are then considering *level logic*, where voltage levels are used to signify logic variables. In this introduction we shall use *positive logic*, which is a convention that a HIGH voltage level, usually +5 V, represents a logic '1'. Accordingly, in positive logic a LOW level, usually 0 V, signifies a logic '0'. The *transfer characteristic* shows the relationship between inputs and outputs, and for an ideal inverter the transfer relationship has only two possible values for the output.

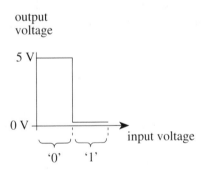

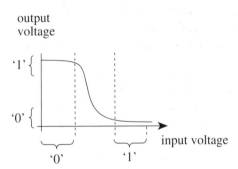

Ideal inverter transfer characteristic   Real inverter transfer characteristic

The slope of the line between the two output states is infinite, which implies that the ideal inverter has an infinite gain in the transition region. This is impossible to achieve in practice. Also, the use of an explicit voltage level to signify a '1' or a '0' is very difficult to achieve. Most circuits use a range of voltage levels to represent logic signals. The transfer characteristic then relates ranges of input and output voltage levels representing '1' and '0'. The transfer characteristic of a real inverter then differs from that of its ideal counterpart.

> *Manufacturers have to guarantee a device's performance limits. These include the electrical and switching characteristics which can be thought of as d.c. and a.c. parameters, respectively.*

For the HIGH state, the logic level most commonly encountered in logic design ranges from 2.0 V to 5.0 V, and for a LOW state the level is between 0 V and 0.8 V to indicate a '0'. This depends on the power supplies used to implement the logic circuit. Any electronic circuit requires a positive and negative power supply, and for a switching circuit these supplies only need to differ in magnitude. The power supply clearly affects the voltage levels used to represent a '1' and a '0'. If two logic technologies use different power supplies then the logic levels in each will be different, and if the two technologies need to be connected together (i.e. a circuit has been designed to take advantage of particular performance characteristics of each logic technology) then this will pose difficulty in *interfacing* (connecting together) the two circuits.

Manufacturers specify *guaranteed output levels* and *specified input levels*. This is a guarantee that if an input is within the range specified for the input levels then the output of the circuit is guaranteed to be within a range specified for the output voltage levels. If you buy a circuit which does not conform to this specification you can return it to the manufacturer for replacement.

If an input voltage level falls outside the range specified for a valid logic level then it naturally becomes invalid and the circuit will not respond correctly. This might occur because noise has corrupted the transmission line connecting two logic gates.

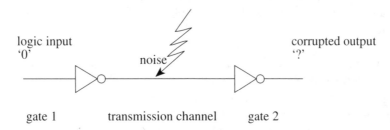

Noise can be induced by electromagnetic radiation from an external source that is of sufficient magnitude to force the voltage level on the transmission channel to become invalid. In reality we actually live in a fog of electromagnetic radiation, and it is unsurprising to find that some gets picked up in logic circuits. The designers of integrated circuits accommodate this by introducing a *noise margin* into a circuit; this is a measure of by how much a valid signal can be corrupted by noise and still be interpreted correctly by the logic circuit. It can be summarised by relating the output levels to the input voltage levels:

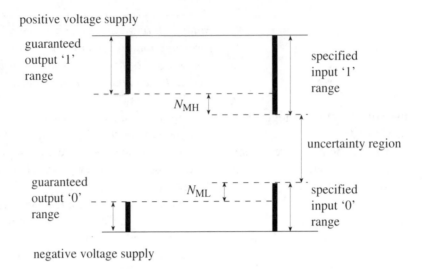

The noise margin for a HIGH level, N_{MH}, is a measure of how much noise can be added to a signal that will still be interpreted as a valid '1'. It is the difference between the minimum level for a valid output '1' and the (specified) minimum level for a valid input '1'. Note that in the HIGH state we are concerned with noise that **reduces** the output level. The noise margin for a LOW level, N_{ML}, concerns noise that **increases** an output level, and is equal to the difference between a valid output '0' and a valid input '0'. N_{ML} is then a measure of how much noise can be added to a LOW logic signal while it is still interpreted correctly by later combinational circuitry. Logic technologies usually specify a noise margin which is equal in both high and low states. There is an uncertainty region between the maximum value for a LOW input and the minimum value for a HIGH input; if a circuit input falls within the uncertainty region then the output is not guaranteed by the manufacturer. This situation should be avoided, since it can result in unreliable circuit operation. It is clearly desirable to have the largest possible noise margin, but some logic technologies can offer only a small noise margin while possessing other attributes.

The logic level might become invalid because of contamination by noise. It might also become invalid because it has been connected to too many outputs. Consider an ideal gate; if the output is a '1' then the output is connected directly to the positive supply. The current available to the output is then that available from the power supply. Circuit inputs consume current and, if we connect the output of one gate to too many subsequent inputs, then there will not be enough current available from the power supply to drive all the subsequent inputs. The whole circuit will then not work. The maximum number of gates that can be connected to the output of a single gate is usually expressed as the *fan-out* of that gate.

> *Some texts refer to 'fan-in' as well, and there are a number of conflicting definitions. Though it is mainly of academic interest only, the most convincing definition of fan-in is that it is the maximum number of logic inputs that a gate can accept. (One definition is that it is the number of input pins!)*

Fan-out can also be explained in terms of valid logic levels. In an ideal gate a HIGH output is connected to the positive supply through a perfect switch. A perfect switch is one that has no resistance when it is switched ON (a short circuit) and infinitely high resistance (an open circuit) when it is switched OFF. When it is ON there is no voltage drop across it since it has no resistance. Logic circuits do not use perfect switches; they use switches that **approximate** to the perfect switch. Since they only approximate their perfect version they have resistance. Their resistance implies that the voltage across them increases with an increase in the current through them. When the voltage drop increases the output voltage falls, eventually to beneath that guaranteed as a valid output. The increase in current will be due to increasing the number of gates connected to the output. If we connect more gates than the specified maximum (the fan-out of the device), the output voltage level will not be valid as specified for the device.

If the difference between logic levels is small then the noise margin is bound to be very small as well. The advantage associated with small differences between logic levels is **speed**, since the logic technology has only to switch the output by a small amount. Speed is usually expressed in terms of *propagation delay*, which is a measure of how long a change in the inputs takes to propagate through to the circuit output. The propagation delay can differ for an output changing from LOW to HIGH or from HIGH to LOW. T_{PLH} measures the time between a change in inputs and the subsequent change in output from '0' to '1'. T_{PHL} is a measure of the time taken for the outputs to change from HIGH to LOW, '1' to '0', in response to a change in the inputs:

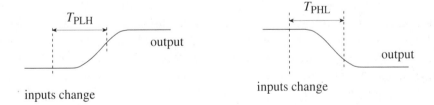

It is advantageous to have the fastest possible switching time. This will usually incur a higher *power consumption*, since this is inherent in a switching process. **Speed** and **power consumption** are often the most important factors in considering different logic technologies. They are often combined in a *speed–power product*, which is formulated by multiplying the worst-case propagation delay (the longest time taken by a device to respond) by the power consumed on average by the device. This can then be used as a performance metric to compare logic technologies.

The main criteria for a logic technology are then the speed it can achieve (the smallest propagation delay), and the power consumed by the device. The secondary performance characteristics are the logic levels, noise margin and fan-out. The secondary criteria might dominate choice in a particular application, since they dominate circuit implementation. All of these factors are a compromise: high speed often incurs high power consumption, while good fan-out usually implies wide noise margins, but since the logic levels are wide the speed is then reduced. Any logic technology then offers a compromise between these factors, and this compromise depends on the switching elements used to implement the logic function, together with implementation criteria. This can be thought of as a switch.

3.2   Switching devices

3.2.1 *The diode*

The simplest form of electronic switch is a *semiconductor diode*. This is essentially a switch that can be turned ON or OFF. Its symbol indicates the direction in which current can pass through it:

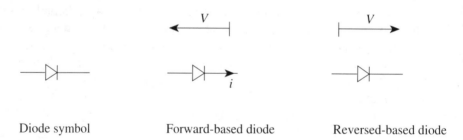

Diode symbol   Forward-based diode   Reversed-based diode

If we apply a voltage consistent with the current direction (the positive connection, the anode, is more positive than the negative connection, the cathode) then current can pass through the device and it is switched ON: in this condition the diode is termed *forward-biased*. If we apply a voltage the other way round (the cathode is more positive than the anode) then no current can flow and the diode is switched OFF: in this condition the diode is termed *reverse-biased*. The *equivalent circuit* is a model of the circuit and for a forward-biased diode it is a short circuit. An equivalent circuit for a reverse-biased diode is an open circuit (since no current can pass). There is actually a **threshold** voltage for a forward-biased diode; the threshold voltage must be exceeded before current can pass, and this is called the *cut-in voltage*. This can be seen from the current–forward-voltage characteristic, which illustrates that a diode does not conduct before the threshold voltage is reached; after the threshold voltage is exceeded, the current increases rapidly.

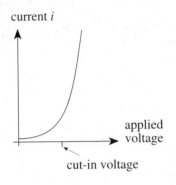

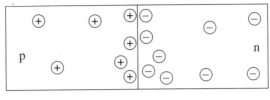

majority carriers, $\ominus$ = electrons, $\oplus$ = holes

Diode *i/V* characteristic

Semiconductor diode

The diode is made by joining two types of semiconductor. One of these is *n-type semi-conductor* (negative type); this is made to have more electrons than normal. The surfeit of electrons is introduced by a process called *doping*. Doping can also cause a deficit of electrons and hence a surfeit of *holes* (essentially the holes left by the absent electrons). If we join these two types of semiconductor we now have a pair of adjacent materials, one with a net negative charge, the other with a net positive charge; this is called a *pn junction*. This is complicated by a depletion region (for more details, see one of the texts referenced at the end of the chapter). Even without any externally applied voltage, electrons in the n-type region will be attracted to the p-type region. There will also be a thermal effect causing electrons to jump the other way. With no externally applied voltage the current generated by thermal effects will balance the current generated by charge effects. When the diode is forward-biased, the potential of the p-type region is raised, and so more electrons will migrate to the p-type region, causing current to flow. When the potential between the p-type and n-type regions is below the cut-in voltage, it is insufficient to attract electrons in large numbers, and no net current flows through the diode.

> *The current is actually exponentially related to the forward bias voltage and thus can increase dramatically. There is a limit to this current since diodes can break. Also, the reverse breakdown voltage specifies a reverse bias sufficiently large for the diode to pass current in the opposite direction.*

Detailed examination of a diode's performance can be found in texts referenced at the end of the chapter (p. 83). The function of a diode in this text is as a switching circuit; we want it to switch ON and OFF. A diode can turn ON if forward-biased and will turn OFF if reverse-biased. This is equivalent to applying a logic '1' and '0':

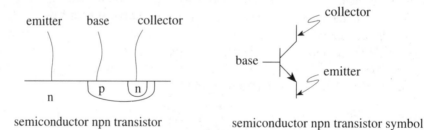

Diode switched on   Diode switched off

3.2.2 Bipolar transistors

A diode has limited functionality in logic circuits, and the *transistor* plays a more important role, primarily because it can operate as an *amplifier*. A transistor is a three-terminal device, and there are two forms:

(a) bipolar – conduction uses both holes and electrons (as in a diode);
(b) unipolar – conduction uses one type of carrier only, either holes or electrons.

A bipolar transistor can be made by forming two regions of n-type semiconductor with one region of p-type. The connections to the n-type regions are termed the *emitter* and the *collector*, and the p-type region is called the *base*. This gives an *npn transistor*:

semiconductor npn transistor   semiconductor npn transistor symbol

The diode between the *base–emitter junction* effectively controls the operation of the device. If the base–emitter junction is forward-biased then electrons are attracted from the emitter into the base and are swept into the collector. Forward biasing the base–emitter junction then causes a collector current to flow through the transistor. If the base–emitter junction potential is below a threshold value then there will be a small current from base to emitter because of charge effects (electrons attracted into the p-type) and a small current from emitter to base (from electrons migrating due to thermal effects). If the base–emitter junction is not forward-biased then there will be no net current, since the thermal effects will balance the charge effects and there will then be no collector current. We can use the base–emitter voltage to switch the collector current on or off. When we introduce a power supply for the collector current, together with a resistor (essentially to limit the collector current), we obtain a circuit that can be used as a switch:

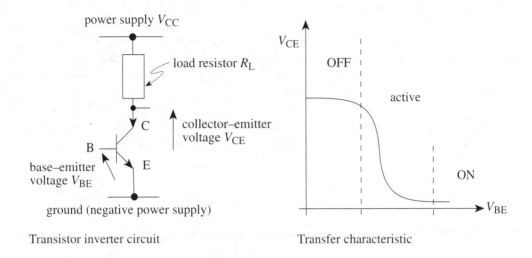

Transistor inverter circuit   Transfer characteristic

The collector current will vary with the base–emitter voltage which in turn causes the collector–emitter voltage to vary. As we forward bias the base–emitter junction the collector current increases and so the collector voltage falls. When the transistor is turned OFF there is no collector current. The transistor is OFF and the collector is connected to the positive voltage supply, so V_{CE} is HIGH. If the base–emitter voltage, V_{BE}, is above a threshold value then the increase in collector current implies a reduction in V_{CE}, which reduces until V_{CE} is LOW. The collector–emitter voltage is the reduction from the power supply voltage by the voltage drop across the load resistor:

$$V_{CE} = V_{CC} - I_C R_L$$

An increase in the collector current I_C implies a voltage drop across the load resistor R_L, causing V_{CE} to be lower than the power supply V_{CC}. There are three named regions associated with the transistor's behaviour; these are marked on the transfer characteristic, which shows the change in V_{CE} with V_{BE}, and are:

(a) the **OFF** region where the collector current is zero;
(b) the **active** region where the collector voltage varies rapidly with changes in V_{BE};
(c) the **ON** region where the collector current is large.

In the active region the slope of the characteristic implies that a small change in V_{BE} will cause a large change in V_{CE}. This is the performance of an amplifier, and the transistor is operated in this region in linear circuits (there is a straight-line approximation in this region of the characteristic). In logic circuits we will be concerned mainly with two states, the ON state and the OFF state. This circuit operates as a logic inverter, since when V_{BE} is HIGH (V_{BE} = '1') then V_{CE} is LOW (V_{CE} = '0'). Conversely, the output is HIGH, '1', when the input, V_{BE}, is LOW and the transistor is switched OFF.

3.2.3   *Field-effect transistors*

There are also *field-effect transistors*, FETs, which have one type of charge carrier controlled by the doping process, either electrons or holes, but not both, as in a bipolar transistor. The technology used to construct these transistors is *metal oxide silicon* (MOS), and the transistors are often known as MOSFETs. There are two types of transistor, consistent with the two types of carrier. One type uses holes and is positive MOS, or PMOS; the other type uses electrons and is negative MOS, or NMOS. FETs again have three terminals but the names on the connections differ from bipolar technology:

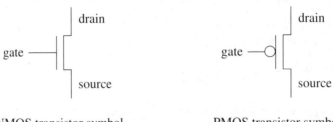

NMOS transistor symbol   PMOS transistor symbol

There are actually many different types of FETs. We are using FETs here within logic technology and it is easiest to think of FETs as **voltage-controlled switches (resistors)**. For both FET transistors the *drain source current*, I_{DS}, is controlled by the *gate source voltage*, V_{GS}. An NMOS transistor switches ON (the *drain current*, I_D, is positive) for positive V_{GS}. If V_{GS} is less than the threshold then the transistor is OFF (I_D is zero). Conversely, a PMOS transistor is ON when V_{GS} is less than a threshold value (and I_D then increases), when V_{GS} is greater than the threshold the transistor is OFF and I_D is zero. This is why an inverter symbol is included within the PMOS transistor symbol. A similar circuit can be developed as with the bipolar transistor to give an inverter; here the inverter is based on an NMOS transistor:

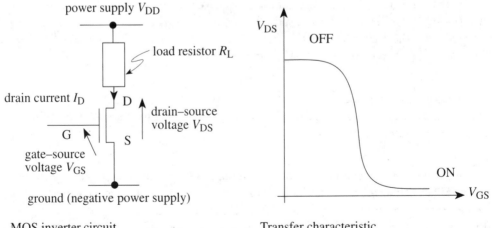

MOS inverter circuit   Transfer characteristic

The power supply is called V_{DD} in MOS technology. Again the circuit is that of an inverter, since if an input is applied between the gate and source then a HIGH input switches the transistor ON and the output is LOW, whereas a LOW input (one lower than the threshold to switch on the NMOS FET) switches the transistor OFF and the output is HIGH, connected to V_{DD}. MOSFETs have major advantages as a logic technology. When the transistor is switched ON the impedance is low (of the order of 100s of ohms), whereas when it is switched OFF the impedance is exceedingly large (of the order of 100s of gigaohms). This is currently the closest possible approximation to an ideal electronic switch.

A bipolar transistor can be thought of as a current-controlled current source (since I_C can depend mainly on the base current I_B when V_{CE} is sufficiently high), whereas an FET is a voltage-controlled current source (since I_D can depend on V_{GS} when V_{DS} is sufficiently large).

The high impedance of a MOSFET can introduce susceptibility to *static damage*. We can generate a potential, via static electricity, in excess of 5 kV just by walking across a carpet. When we touch a chip input, the resistance of the air gap between the chip input and our finger is about 1 GΩ. If the chip input circuit has a small impedance compared with 1 GΩ, then the potential will be dropped across the air gap and we will see a spark. If the impedance of the chip input is large compared with the air gap (the, say, 9 GΩ impedance of a MOSFET is clearly large compared with 1 GΩ), then the potential will be dropped across the chip input circuit. For a MOSFET this is actually larger than an internal *breakdown* voltage, i.e. the maximum voltage that can be applied across parts of the device. If we exceed it we will damage the device. This is why CMOS devices are known as *static-sensitive* and, even though modern designs have protection built into circuits, it is often prudent to avoid handling MOS circuits directly (without connecting yourself to 0 V to avoid static potential).

3.3 Logic technologies

3.3.1 *Complementary metal oxide silicon (CMOS)*

CMOS evolved from combining NMOS and PMOS transistors. An NMOS inverter can be made by connecting the drain of an NMOS transistor to the positive voltage supply using a resistor, as described in the previous section. Resistors actually require a large amount of space in integrated circuits. To save space, the resistor is replaced with an NMOS transistor which is permanently switched on. The ON resistance of the NMOS transistor can be carefully controlled and the NMOS transistor is physically smaller than an equivalent resistor.

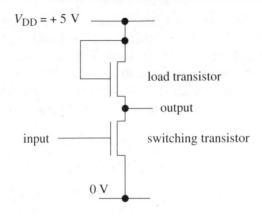

NMOS inverter

The load transistor is permanently switched ON by connecting its gate input to the positive power supply. When the NMOS switching transistor is switched ON the output is connected through it to ground. When the switching transistor is OFF, the output is connected HIGH through the load transistor. In principle the circuit operates as an inverter. One drawback is that when both transistors are ON then the output is effectively derived from a potential divider circuit. If both transistors have the same on-resistance, the output will be approximately half the power supply voltage, rather than ground, $0\,V$. Designers then control the on-resistance of the load transistor to ensure that a LOW output is close to $0\,V$, the negative voltage supply. This is called *ratioing*, where we modify the physical dimensions of the load resistor (by modifying the aspect ratio of width to length) to increase its on-resistance.

There is also a PMOS inverter, which again avoids using resistors by using a PMOS transistor which is again permanently switched on:

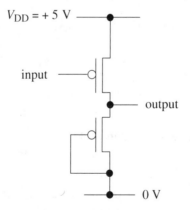

PMOS inverter

Positive MOS depends on holes as the majority carrier. This implies that PMOS transistors must be larger than their NMOS counterparts, owing also to the decreased mobility of holes when compared with that of electrons. Since smaller size implies smaller capacitances and hence faster circuits (and other practical difficulties), NMOS dominated before CMOS. The difficulties associated with ratioing motivated development which centred on using both NMOS and PMOS FETs. Since these switch ON for inverted inputs, their combination is called *complementary MOS*; this therefore avoids the problem of ratioing.

A CMOS inverter has one transistor to switch the output HIGH for a LOW input. This requires a PMOS transistor. A LOW output is caused by a HIGH input, which requires a NMOS transistor.

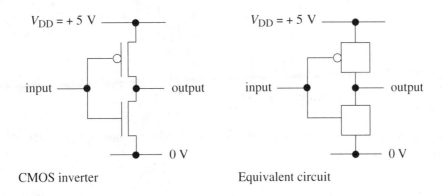

CMOS inverter   Equivalent circuit

The equivalent circuit for a CMOS inverter is a pair of switches on which the input acts in a complementary way. The CMOS inverter is at present the closest possible approximation to the ideal inverter.

> *If we compare the CMOS inverter with the ideal inverter in the diagram at the beginning of this chapter, the FETs just replace the switches. For logic gates (such as NAND and NOR), CMOS circuits actually combine the output stage within the switching function rather than use a combinational logic function as shown.*

The operation of the CMOS inverter can be described by a table showing which transistors are OFF or ON in each logic state.

Input	NMOS	PMOS	Output
'0'	OFF	ON	'1'
'1'	ON	OFF	'0'

A 2-input CMOS NAND gate is designed to give a LOW output when both inputs are HIGH. This can be achieved by two NMOS transistors in series, N1 and N2, connecting the output LOW. Conversely, if either or both inputs are LOW then the output should be connected HIGH. This can be achieved by a pair of PMOS transistors, P1 and P2, which are connected in parallel to the positive voltage supply:

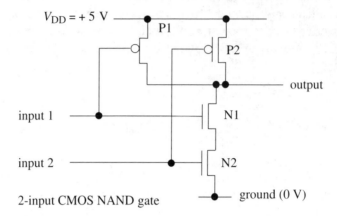

2-input CMOS NAND gate

The activity of the device can be described in terms of which transistors are ON or OFF according to the values of the inputs.

Input 1	Input 2	P1	P2	N1	N2	Output
'0'	'0'	ON	ON	OFF	OFF	'1'
'0'	'1'	ON	OFF	OFF	ON	'1'
'1'	'0'	OFF	ON	ON	OFF	'1'
'1'	'1'	OFF	OFF	ON	ON	'0'

It can be seen that the parallel PMOS transistors serve to connect the output HIGH if either input is LOW. This is actually an OR structure; if input 1 OR input 2 is LOW then the output is HIGH. The NMOS transistors serve to pull the output low when they are both turned ON. This is an AND structure. If input 1 AND input 2 are HIGH then the output is LOW. This is an illustration of de Morgan's law:

$$\overline{A \cdot B} = \overline{A} + \overline{B}$$

The equivalent circuits for a NAND gate sourcing HIGH and LOW outputs concern only the transistors that are switched ON. The large impedance of a MOSFET that is OFF implies that they effectively disappear from the circuit. Using dotted lines to indicate the parts of the circuit which are switched OFF, the circuit for a CMOS NAND gate when sourcing a HIGH or a LOW output becomes

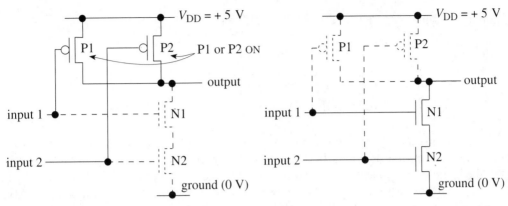

CMOS NAND gate sourcing a HIGH output   CMOS NAND gate sourcing a LOW output

The CMOS 2-input NAND gate has two circuits, one of which provides a NAND gate for LOW inputs (the PMOS transistors); the other is a NAND gate for HIGH inputs (the NMOS transistors). This is reflected in other CMOS gates. The 2-input CMOS NOR gate provides an output that is LOW when either input is HIGH and an output that is HIGH when both inputs are LOW. The parallel OR structure is used to give a LOW output. The series AND structure is again used to provide a HIGH output when both inputs are LOW.

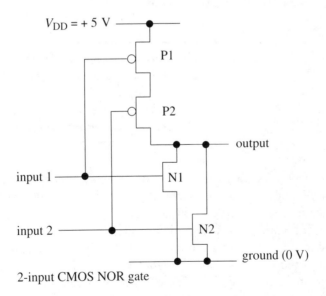

2-input CMOS NOR gate

The activity of the device can again be described in terms of which transistors are ON or OFF according to the values of the inputs:

Input 1	Input 2	P1	P2	N1	N2	Output
'0'	'0'	ON	ON	OFF	OFF	'1'
'0'	'1'	ON	OFF	OFF	ON	'0'
'1'	'0'	OFF	ON	ON	OFF	'0'
'1'	'1'	OFF	OFF	ON	ON	'0'

The main advantage of CMOS technology is that it has very low power consumption. This occurs when the device is not switching and is termed the *static power dissipation.* That the static power is low is intimately related to the closeness of FETs to the ideal gate. The large OFF resistance implies very low power consumption. Disadvantages of MOS technology include a poor drive capability for large capacitative loads (the current sourced by MOSFETs can be less than that for bipolar transistors). Also, the power consumption increases with switching speed. The power consumption is actually related to other factors, as follows:

$$\text{power consumption} \propto f \times (V_{\text{supply}})^2 \times C$$

where f is the switching frequency, V_{supply} is the positive voltage supply and C is the capacitance driven. It can be observed that the power supply plays a major role in CMOS performance characteristics. The fan-out of CMOS circuits and the noise margins are good.

> *One advantage of CMOS is that the maximum power supply can range from 3V up to 7V, and this can be used by designers to optimise performance. Note that low-power-supply logic is likely to become very important in the (near) future.*

The major advantages of CMOS are that current versions can achieve the moderate speeds required for general-purpose logic with lower power consumption than other logic technologies. For these reasons, CMOS now dominates much of integrated circuit design. The early CMOS *logic family,* which provided a range of logic functions, was the 4000 series CMOS introduced in 1972. This offered very low power consumption, but with very low-speed operation (about 80 ns propagation delay per gate). This family offered ranges of gates from simple combinational gates to more complex circuits. Though slow, the 4000 series dominated early implementations where power consumption was critical. Developments in CMOS manufacturing technology have led to the present position, where CMOS now dominates implementation; this was achieved by supplanting a famous and popular logic technology that used bipolar transistors, TTL logic.

3.3.2 Transistor–transistor logic (TTL)

TTL is one of the landmarks in digital design. It is a logic technology based on using bipolar transistors as the switching elements. The technology was introduced much earlier than MOS. Texas Instruments introduced the *74 series TTL* logic family in 1964, and it dominated logic implementation for many years; its performance could not be bettered in general-purpose applications, and the range of functions offered within the logic family was very wide. It has now been largely supplanted by MOS technology, which offers a similar performance but with reduced power consumption. The pre-eminence of 74 series TTL is reflected in the 74 series CMOS, which is a CMOS version giving direct replacement for TTL logic. Though there are some advantages in using TTL, it is today finding fewer and fewer applications. However, the development of a TTL gate serves to illustrate how a logic gate was developed to satisfy the performance characteristics of its operation.

3.3.2.1 Development of a TTL NAND gate

The starting point for the development of a TTL gate is *diode logic*. This was an early form of logic, which used only diodes as the switching elements. A 2-input OR gate gives a HIGH output for HIGH inputs, and can be implemented using two diodes in parallel. When both inputs are LOW, both diodes are OFF and the output is connected LOW through the resistor. The circuit then gives a HIGH output if either input is HIGH, and a LOW output when both inputs are LOW, as consistent with the OR function.

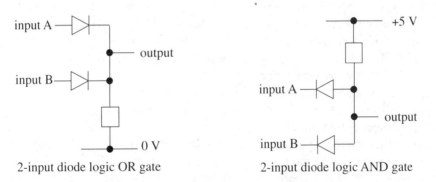

2-input diode logic OR gate   2-input diode logic AND gate

For an AND gate, the diodes are reversed and connected HIGH through a resistor. If either input is LOW then one of the diodes will be ON, so the voltage at the output will be LOW (and equal approximately to the voltage across the forward-biased diode). When both inputs are HIGH, neither diode is forward-biased, and so the output will be connected HIGH through the resistor. The output is then HIGH when both inputs are HIGH, and LOW for any other input combination, which is consistent with the AND function.

One major disadvantage with diode logic is that it is not possible to construct an inverter circuit. Another disadvantage is its poor **fan-out**. Output signals are effectively derived from input signals; the input power provides the output power. A bipolar transistor circuit can be used to improve the fan-out. If we include an inverter circuit after a diode logic AND gate then we will obtain a *diode transistor logic* (DTL) NAND gate (since we have AND followed by NOT, giving NAND).

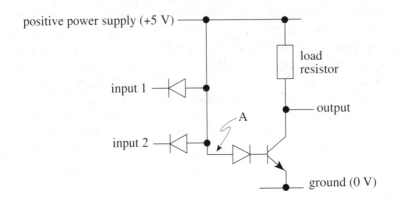

The output is now not connected directly to the inputs, since when the transistor is ON the output is connected LOW to ground through the ON transistor. When the transistor is OFF then the output is connected HIGH through the load resistor. An extra element has been introduced: there is a diode between the diode logic AND gate and the inverter. This is called a *coupling diode*, and it is included to ensure that the transistor is ON when both inputs are HIGH but, more importantly, to ensure that the transistor is OFF when either input is LOW. When either diode is ON, there is a voltage drop across it. The voltage drop across a forward-biased diode is sufficient to turn on a transistor (since the base–emitter junction can be considered as a diode and V_{BE} for an ON transistor is greater than 0.5 V). If the coupling diode were not there, then a LOW input could turn the transistor ON. If both inputs are HIGH, then the positive voltage supply is applied to the base of the transistor and it is again ON. If the coupling diode was not included then the transistor would be ON all the time. The coupling diode ensures that the transistor is OFF if either input is LOW. If either input is LOW then the voltage at point A is insufficient to turn on both the coupling diode and the base–emitter junction of the transistor.

Inclusion of a further coupling diode, so that there are two diodes between point A and the transistor, helps to improve the **noise margin**. If the input is LOW then the voltage at point A is the input LOW voltage, plus the voltage across the forward-biased input diode. The voltage at point A will then rise with the input voltage. If there is a single coupling diode then we can increase the input voltage to the cut-in voltage of the coupling diode, plus the threshold voltage of the transistor base–emitter junction. If there are two coupling diodes then we can raise the input voltage further and the circuit will continue to operate correctly. With two coupling diodes, we can increase the input voltage more than with a single coupling diode, and still obtain a valid output, and so the noise margin is increased.

The major restriction on the DTL NAND gate concerns its response time. If a capacitive load is attached to the output then the capacitance will charge up when the output is HIGH and discharge when the output is LOW. When the output is LOW the capacitance discharges through the ON transistor, which offers low impedance. When the transistor switches OFF the output capacitance will charge up through the load resistor. The *time constant* measures the rate in increase of the output voltage and is equal to the product of the load resistance and the output capacitance. If either of these is large then the output voltage will take an appreciable time to reach the HIGH output state. When the output switches LOW, the output voltage returns LOW quite fast, since the time constant associated with the ON transistor can be smaller.

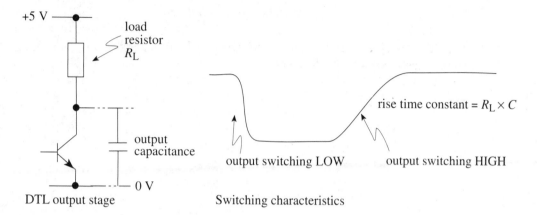

DTL output stage   Switching characteristics

Switching speed is improved in a TTL NAND gate, which uses a *totem pole* output stage. The totem pole comprises two transistors, one to switch the output HIGH, the other to switch the output LOW. The transistor pulling the output HIGH is termed an *active pull-up* circuit and it replaces the load resistor in the DTL NAND gate.

> *It is possible that logic connections are more inductive than capacitive. However, it is clearly an improvement to use the separate transistors in a totem pole output stage.*

The full TTL NAND gate does not use the diodes in the DTL input stage. An npn bipolar transistor can be modelled as two diodes placed back-to-back. This is because a diode is formed by a pn junction and there are two pn junctions in the bipolar transistor. The diode AND circuit requires two inputs, and this is achieved by connecting an extra input to the emitter and the transistor is then termed *multi-emitter*. The multi-emitter transistor replaces the diode AND circuit and the coupling diode in the DTL NAND gate.

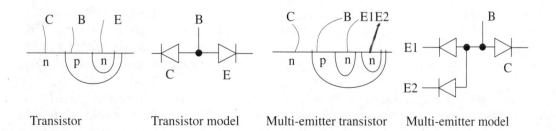

Transistor   Transistor model   Multi-emitter transistor   Multi-emitter model

> *Modelling a transistor using diodes is actually the*
> *basis of the Ebers–Moll transistor model, which*
> *further includes an ideal current sources in*
> *parallel with each diode.*

The totem pole output stage and the multi-emitter transistor are added in to the DTL NAND gate to provide a full TTL NAND gate:

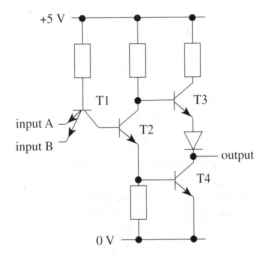

2-input TTL NAND gate

The function of the circuit centres on switching the output HIGH (by switching on transistor T3) or switching the output LOW (by switching on transistor T4). T3 and T4 should not both be ON at the same time. Switching naturally centres on the input values, which we will assume are HIGH (+5 V) or LOW (0 V). If either input is LOW then the base–emitter voltage of transistor T1 is sufficient to switch T1 ON. When T1 is ON the collector-emitter voltage is LOW and so the collector voltage is approximately 0 V. The collector of T1 feeds the base of T2, and, if the base of T2 is LOW at about 0 V, then transistor T2 is switched OFF. If T2 is OFF then there is no collector current through T2, so the collector of T2 will be connected to the positive rail via a resistor. The collector of T2 feeds the base of T3, so the base of T3 will be HIGH, turning it ON. The output will then be HIGH because with T3 ON the output is connected, through T3 and a diode, to the positive voltage supply. If either input is LOW then T1 is ON, turning OFF T2, which causes T3 to switch ON, giving a HIGH output, and T4 is also OFF as required. Note that T2 is often called a *phase splitter*, since its function is to switch T3 or T4 in a complementary manner (T3 is ON when T4 is OFF).

Conversely, if both inputs are high then the output is LOW. If both inputs are HIGH then T1 would appear to be OFF, since the base–emitter junction is not forward-biased. In this case T1 operates in the *inverse active* mode, where the collector current comes from the base current and the emitter current. It is termed inverse active because it is operating as a transistor which has been inverted (the emitter has become the collector and vice versa). In the usual active mode the emitter current derives from the collector and the base currents. In the inverse active mode we turn the device around (even though it is asymmetric in physical size it is functionally symmetric since it is an npn device) to derive a collector current from the emitter current injected by HIGH inputs, together with a base current. This implies that current is injected into T2, switching it ON (the voltage at the base of T2 is HIGH also) and, since T2 is ON, its collector voltage falls, switching OFF T3, while the emitter voltage is sufficient to turn T4 ON. The diode is included to ensure that T3 is OFF when T4 is ON. This functions in a manner similar to the coupling diode in the DTL NAND gate, since if the output is LOW then the voltage at the base of T3 would be just sufficient to turn it ON (since T4 is ON). The diode is then included to ensure that when T4 is ON, T3 is OFF, since the potential at the base of T3 is then insufficient to turn on both T3 and the diode.

Input A	Input B	T1	T2	T3	T4	Output
'0'	'0'	ON	OFF	ON	OFF	'1'
'0'	'1'	ON	OFF	ON	OFF	'1'
'1'	'0'	ON	OFF	ON	OFF	'1'
'1'	'1'	Inverse active	ON	OFF	ON	'0'

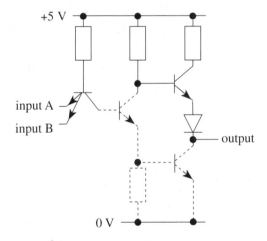

TTL NAND gate giving HIGH output

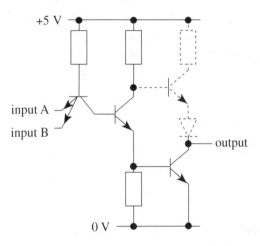

TTL NAND gate giving LOW output

Inductive input leads can cause negative input voltages, and so diodes are often used to prevent negative gate input voltages. The protection diodes are connected between the inputs and 0V to be forward-biased for a negative input, thus ensuring that the gate input voltage never falls below the diode's cut-in voltage.

It was stated earlier that TTL is now becoming obsolescent. This is because even though it can achieve medium speed, the power consumption is excessive when compared with modern CMOS. The advantages of CMOS are little when the output load has a large capacitance and it is being driven at medium speed. The power consumption is reflected in the number of bipolar transistors required by the circuit, which are a poorer match to an ideal switch than CMOS. The development of the TTL NAND gate does, however, serve as an excellent vehicle to see how logic designers can include important performance characteristics in the design of a device.

3.3.2.2 *TTL variants*

The basic TTL gate is of historic interest only now. Early variants included a low-power version, which was denoted 74LXXX, and reduced the power consumption by using larger resistors, which concurrently reduced operating speeds. Speed was improved in *Schottky TTL*, which avoided transistor *saturation*. When the transistor is switched on there is excess base current, which results in stored charge in the device. When it is switched fully on, with this excess base current, the collector–emitter voltage is low; this is termed saturation. When we want to switch the transistor off, we must remove this excess stored charge within the transistor, which takes time. Operating speeds can be improved by avoiding saturation; this involves designing a circuit that prevents the collector–emitter voltage falling too far. Saturation can be analysed using the diode model of a transistor:

| Transistor diode model | Transistor in saturation | Transistor (ON) plus Schottky diode | Schottky transistor |

If the transistor is ON then the base–emitter junction is forward-biased. Saturation implies a low collector–emitter voltage, which in turn implies that, in saturation, the base–collector junction is forward-biased. If the base–collector voltage is reduced then the transistor leaves saturation. Schottky TTL uses a *Schottky barrier diode* (SBD), which has a low cut-in voltage that fixes the base–collector voltage to **less** than that for a saturated transistor. By connecting the Schottky diode at the base–collector junction we clamp the base–collector voltage to the cut-in voltage of the diode and it can increase no further.

> *Other properties of the Schottky diode include a fast switch-on time owing to the small charge storage capacity. Schottky diodes are quite simple to build in bipolar technology – a special base contact straddles the base–collector junction to provide the diode.*

An alternative interpretation is that the Schottky diode diverts the excess base current from the base–collector junction. On inclusion of the Schottky diode, the transistor becomes a *Schottky transistor*. Logic circuits made from Schottky transistors, denoted **74S**XXX, differ from those of the basic TTL gate. Since the transistors cannot enter saturation, the operation of the device is faster, because it operates within the edges of the active region.

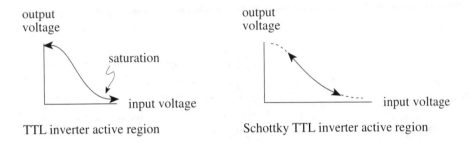

TTL inverter active region   Schottky TTL inverter active region

The most popular of all TTL families was the low-power Schottky TTL family, denoted 74LSXXX, which was a popular, general-purpose logic family for nearly 20 years. LSTTL was a good all-round performer. It could achieve medium speed with medium power consumption. The LS family also offered a wide range of chips, from basic NAND gates to very complex integrated circuits. The advanced low-power Schottky TTL, 74ALSXXX, was introduced later and achieved roughly twice the speed with half the power consumption of LSTTL. A much faster version called fast TTL, 74FXXX, indeed offered very high speed, but with increased power consumption and with a much narrower range of devices than LSTTL. The enduring popularity of 74 series logic was reflected in the choice of CMOS to implement the 74 series. Two of the main CMOS versions of 74 series logic are high-speed CMOS, 74HCXXX, and a high-speed CMOS which offers direct replacement for LSTTL, 74HCTXXX. The CMOS versions of 74 series logic were introduced just after the introduction of the 74F and the 74ALS series. Since CMOS gives equivalent performance with lower power consumption and has started to dominate implementation, this is perhaps why there is only a limited range of devices in 74F and 74ALS technologies.

3.3.3 *Emitter-coupled logic (ECL)*

Emitter-coupled logic is perhaps the oldest logic technology still in current use. It is a bipolar technology with transistor-based circuits. These circuits avoid transistor saturation by using a circuit known as a long-tailed pair in which the emitters of two transistors are connected together, hence the name ECL.

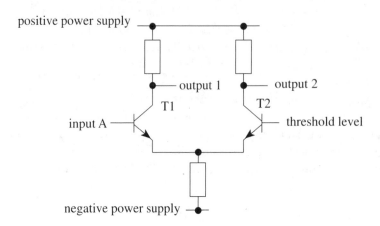

The long-tailed pair circuit operates as a differential amplifier. In terms of a logic circuit, if the input is greater than the threshold level, then the transistor T1 will switch ON, ensuring that the emitter voltage is high enough to force the base–emitter voltage of T2 to be LOW, switching T2 OFF. Conversely, if input A is less than the threshold level, then T2 is switched ON, forcing the voltage of the coupled emitters to be too high to turn on T1, and so T1 is OFF. In order to make a logic function we can include another transistor in parallel with transistor T1. We will then switch the left-hand side on if either or both inputs are higher than the threshold level. If both inputs are lower than the threshold level then the right-hand side is switched on and the left-hand side is off. This gives a logic functionality equivalent to an OR/NOR structure:

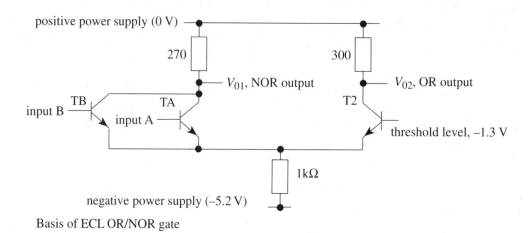

Basis of ECL OR/NOR gate

Note that the power supplies are $0\,V$ for the positive power supply and $-5.2\,V$ for the negative power supply. This in turn implies that the logic levels are negative. In ECL the logic levels are:

$$\text{HIGH: `1'} = -0.8\,\text{V} \qquad \text{LOW: `0'} = -1.6\,\text{V}$$

Also, in ECL the transistors are slightly different, in that they switch on for a slightly higher base–emitter voltage of 0.8 V, compared with the switch-on voltage in TTL; the switch-on voltage $V_{BE}(ON) = 0.8\,\text{V}$.

> *ECL can operate with power supplies other than those indicated, since the power supplies only need to differ in magnitude. Negative power supplies are actually used for practical considerations; in ECL, the high speed and low noise margins cause correct connection termination to be vital.*

In this circuit, if input A or input B is high, –0.8 V, then either or both of the input transistors is ON. This implies that at least one of the left-hand side transistors is ON, and, since $V_{BE}(ON) = 0.8\,\text{V}$, the emitter is 0.8 V below the base and is then –1.6 V. This emitter voltage is that of the right-hand-side transistor, T2, and since the threshold level is –1.3 V, $V_{BE}(T2) = 0.3\,\text{V}$, which forces T2 to be OFF. If either or both inputs is HIGH then the left-hand side is ON, forcing T2 to be OFF. The current through the emitter resistor is then

$$i_e = [-1.6 - (-5.2)]/1.2 \times 10^3$$

$$= 3\,\text{mA}$$

The voltage drop across the collector resistor is then the output voltage V_{o1}:

$$V_{o1} = i_e \times 270 = -0.8\text{V}$$

In this situation T2 is OFF, so V_{o2} is connected to 0 V (the positive supply).

Conversely, if both inputs are LOW, –1.6 V, then both left-hand-side transistors TA and TB are OFF, since T2 is ON, the emitter voltage is then 0.8 V below the threshold level and so $V_{BE}(TA \text{ and } TB) = 0.5\,\text{V}$, which is less than the level to switch these transistors on.

If both inputs are LOW and T2 is ON then the emitter current is given by

$$i_e = [-2.1 - (-5.2)]/1.2 \times 10^3$$

$$= 2.6\,\text{mA}$$

The voltage drop across the collector resistor is then the output voltage V_{o2}:

$$V_{o2} = i_e \times 300 = -0.8\text{V}$$

In this situation T1 is OFF so $V_{o1} = 0\,\text{V}$. This operation can be summarised as follows:

Input A	Input B	TA	TB	T2	V_{o1}	V_{o2}
'0'	'0'	OFF	OFF	ON	0 V ('1')	−0.8 V ('0')
'0'	'1'	OFF	ON	OFF	−0.8 V ('0')	0 V ('1')
'1'	'0'	ON	OFF	OFF	−0.8 V ('0')	0 V ('1')
'1'	'1'	ON	ON	OFF	−0.8 V ('0')	0 V ('1')

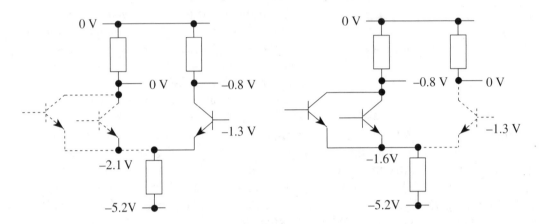

ECL gate for LOW inputs   ECL gate for either or both inputs HIGH

These output voltage levels are different from the input voltage levels. For output voltage levels '0' = −0.8 V and '1' = 0 V then V_{o1} and V_{o2} can be seen to implement NOR and OR respectively. The output levels then need to be corrected to be the same as the input voltage levels. We need to subtract 0.8 V from each output voltage level to return to the original input voltage levels. This can be arranged using a transistor connected as an *emitter follower*:

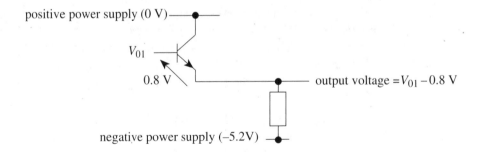

The required drop of 0.8 V is then used to switch the output stage ON. The emitter follower circuit has several practical advantages. It restores the logic levels but, since it has high current gain (current is effectively drawn from the power supply), fan-out is improved considerably. The same circuit is used to correct both the OR and the NOR outputs.

Note that the ECL circuit avoids saturation since when the left-hand side is ON,

$V_{o1} = -0.8\,\text{V}$ and $V_e = -1.6\,\text{V}$, so $V_{ce1} = 0.8\,\text{V}$, which implies that when the left-hand side is ON then neither TA nor TB reaches saturation. Conversely, when T2 is ON then

$$V_{o2} = -0.8\,\text{V} \text{ and } V_e = -2.1\,\text{V}, \text{ so } V_{ce2} = 1.3\,\text{V}$$

Clearly, when any of the transistors are ON, they cannot reach saturation. The circuit is actually the fastest logic technology currently available and can achieve propagation delays of less than 1 ns. This is because the circuit avoids saturation, and because we are switching between logic levels that are very close together. Since the logic levels differ by only a small amount the noise margin is reduced and designers are very careful when implementing circuits in ECL. Note that the emitter–follower output stages increase fan-out but also increase the number of transistors. A true ECL circuit is even more complicated, requiring even more transistors, and so the power consumption is high. This is the (inevitable) price paid for inherently high-speed operation.

3.3.4   Bipolar CMOS (BiCMOS)

BiCMOS is a combination of bipolar and CMOS logic, designed to take advantage of the best performance characteristics of each. It is self-evident that manufacturing costs are reduced by making integrated circuits increasingly smaller. Miniaturisation also incurs higher capacitances associated with small physical size and CMOS is poor at driving capacitative loads but with good (small) static power dissipation. On the other hand, bipolar logic can drive capacitative loads but with increased power consumption. If the two technologies are combined, using CMOS to implement the switching circuit and a bipolar output stage to drive the load, then we utilise the best performance characteristics of each technology.

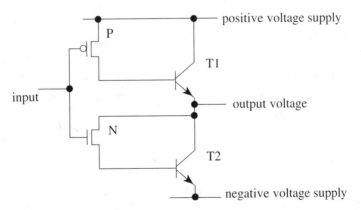

Basic BiCMOS inverter circuit

For a basic BiCMOS inverter circuit, a HIGH input causes the PMOS transistor to be OFF and the NMOS transistor to be ON. This in turn causes the pull-up transistor T1 to be OFF and T2 to be ON. A HIGH input then results in a LOW output. Conversely, a LOW input switches the PMOS transistor ON, and the NMOS transistor OFF, and hence T1 ON and T2 OFF. A LOW input then causes a HIGH output.

In the static case it is not clear why the transistors switch ON at all. A LOW output appears to switch on a bipolar transistor (T2). The circuit operation actually derives from its switching properties. If the input voltage changes from LOW to HIGH then the output voltage is initially HIGH (and will go LOW). When the input goes HIGH this switches the NMOS transistor ON, connecting the output (which is initially HIGH) to the base of the lower transistor T2. The output is then sufficient to turn T2 ON, pulling the output LOW. The output voltage falls to a value just below the base–emitter junction voltage of T2, and stays there.

BiCMOS actually achieves low power consumption because no large currents are switched within a device. The bipolar transistors are turned either just ON or just OFF.

The circuit actually operates in a manner similar to Schottky logic. The NMOS transistor is connected between the base–collector junction of T2, and so when the NMOS transistor is ON it clamps the base–collector voltage of T2 to prevent the transistor reaching saturation. The output voltage is then equal to the base–emitter voltage of T2, minus the (small) voltage drop across the NMOS transistor. The upper circuit is also similar to a Schottky circuit since when the PMOS transistor is ON it clamps the base–collector junction of T1, again ensuring that T1 does not reach saturation. When the output switches from HIGH to LOW the PMOS transistor switches ON. This switches on T1 to pull the output HIGH. The output will rise to just below the power supply voltage, minus the base–emitter voltage across T1.

In practice BiCMOS circuits are more complex than this. This is because charge stored on transistor bases needs to be moved before the device can change state. When driving small capacitative loads CMOS can outperform BiCMOS. When driving large capacitative loads, BiCMOS can be two to five times faster than CMOS. For example, in a large computer memory circuit, CMOS can achieve propagation delays of about 13 ns, whereas BiCMOS is faster at 8 ns. The major disadvantage is production complexity and hence cost. Owing to its drive capability, BiCMOS has dominated *data conversion* products where we connect logic circuits to the outside world. It is now finding increasing use in logic circuits in general.

3.4   Comparison of logic circuits

The main criteria for comparing logic families and technologies are power consumption and speed. The various logic families considered have a variation in performance characteristics according to differing elements of the logic technology and the variants within the logic technology. The power supply chosen for operation can also affect performance characteristics greatly, as in 74 series CMOS. The families can be summarised by point measures for elements in the logic family, together with an overall view treating logic technologies in general:

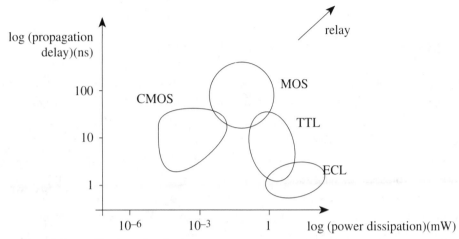

Comparison of logic technologies

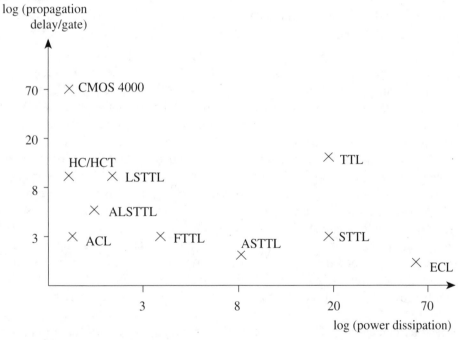

Comparison of logic families

There are many other important factors not yet mentioned, such as cost and availability. Also, connecting together a chip is not the whole story for implementing circuits: *shielding* to protect inputs from noise plays an important part in some designs, while other circuits require interfacing between two logic technologies; these factors will be considered in Chapter 8. Databook specifications on logic gates can be interpreted in the light of these general criteria.

3.5 Specimen CMOS and ECL datasheets

The following datasheets are reproduced with kind permission of Motorola Inc. from the FACT data book describing Motorola advanced CMOS logic. The CMOS datasheet describes a hex inverter chip implemented as pure CMOS (the AC04) or as one directly compatible with TTL (ACT04). There is a wide range of possible power supply values (especially the AC04), which implies a wide range of performance characteristics since the rise time is expressed in nanoseconds per volt. The voltage input and output levels are compatible between AC and ACT logic, but only ACT logic is directly compatible with TTL consistent with its power supply voltage range (4.5 to 5.5 V). These CMOS devices offer, essentially, a robust all-round performance and are therefore well suited to general-purpose designs.

The ECL datasheet shows an OR/NOR gate with its high-speed, nanosecond response time. This chip has three power supply connections: one for the positive power supply, one for negative and the other for ground. The MECL 100K gate offers much faster response time than an earlier MECL 10K series, with the same power dissipation.

The DC voltage characteristics differ between AC and ACT at the input stage only. ACT accepts, as a guaranteed limit, input voltages which are the same as for TTL, namely that $V_{IH}(min)$ is 2.0 V and $V_{IL}(max)$ is 0.8 V whereas for AC the guaranteed limits are $V_{IH}(min)$ is 3.15 V and $V_{IL}(max)$ is 1.35 V (for V_{CC} = 4.5 V). This is consistent with the design aims of the AC and ACT families. The output voltage limits are specified to be the same for both AC and ACT, as are the DC current characteristics. A wider range of limits for the input and output voltages reflects the improved fan-in of CMOS. For ECL (at 25°) $V_{IH}(min)$ is –1.13 V and $V_{IL}(max)$ is –1.48 V and $V_{OH}(min)$ is –0.98 V and $V_{OL}(max)$ is –1.63 V (at both input and output $V_H(max)$ is –0.81 V and $V_L(min)$ is –1.95 V. This implies that the noise margin for ECL is poor since the logic levels are close together.

The propagation delay of ACT is specified as typically TPLH = 4.0 ns and TPHL = 3.5 ns (for VCC = 5 V), and the limit is less than 10 ns for both AC and ACT. The limit for ACT is actually specified as being just larger than that for AC logic. Both families are considerably slower than ECL, which specifies a maximum propagation delay of just over 1 ns.

The speed associated with ECL has the penalty of larger power consumption; note that the power-supply current for ECL is quoted in units of mA, whereas the quiescent power-supply current for AC and ACT is quoted in units of mA.

 **MOTOROLA**

MC74AC04
MC74ACT04

HEX INVERTER

Hex Inverter

- Outputs Source/Sink 24 mA
- 'ACT04 Has TTL Compatible Inputs

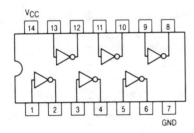

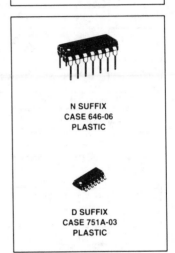

N SUFFIX
CASE 646-06
PLASTIC

D SUFFIX
CASE 751A-03
PLASTIC

MAXIMUM RATINGS*

Symbol	Parameter	Value	Unit
V_{CC}	DC Supply Voltage (Referenced to GND)	−0.5 to +7.0	V
V_{in}	DC Input Voltage (Referenced to GND)	−0.5 to V_{CC} +0.5	V
V_{out}	DC Output Voltage (Referenced to GND)	−0.5 to V_{CC} +0.5	V
I_{in}	DC Input Current, per Pin	±20	mA
I_{out}	DC Output Sink/Source Current, per Pin	±50	mA
I_{CC}	DC V_{CC} or GND Current per Output Pin	±50	mA
T_{stg}	Storage Temperature	−65 to +150	°C

* Maximum Ratings are those values beyond which damage to the device may occur. Functional operation should be restricted to the Recommended
Operating Conditions.

MC74AC04 • MC74ACT04

RECOMMENDED OPERATING CONDITIONS

Symbol	Parameter		Min	Typ	Max	Unit
V_{CC}	Supply Voltage	'AC	2.0	5.0	6.0	V
		'ACT	4.5	5.0	5.5	
V_{in}, V_{out}	DC Input Voltage, Output Voltage (Ref. to GND)		0		V_{CC}	V
t_r, t_f	Input Rise and Fall Time (Note 1) 'AC Devices except Schmitt Inputs	V_{CC} @ 3.0 V		150		ns/V
		V_{CC} @ 4.5 V		40		
		V_{CC} @ 5.5 V		25		
t_r, t_f	Input Rise and Fall Time (Note 2) 'ACT Devices except Schmitt Inputs	V_{CC} @ 4.5 V		10		ns/V
		V_{CC} @ 5.5 V		8.0		
T_J	Junction Temperature (PDIP)				140	°C
T_A	Operating Ambient Temperature Range		−40	25	85	°C
I_{OH}	Output Current — High				−24	mA
I_{OL}	Output Current — Low				24	mA

1. V_{in} from 30% to 70% V_{CC}; see individual Data Sheets for devices that differ from the typical input rise and fall times.
2. V_{in} from 0.8 V to 2.0 V; see individual Data Sheets for devices that differ from the typical input rise and fall times.

DC CHARACTERISTICS

Symbol	Parameter	V_{CC} (V)	74AC T_A = +25°C	74AC T_A = −40°C to +85°C	Unit	Conditions
			Typ	Guaranteed Limits		
V_{IH}	Minimum High Level Input Voltage	3.0	1.5	2.1 — 2.1	V	V_{OUT} = 0.1 V or V_{CC} − 0.1 V
		4.5	2.25	3.15 — 3.15		
		5.5	2.75	3.85 — 3.85		
V_{IL}	Maximum Low Level Input Voltage	3.0	1.5	0.9 — 0.9	V	V_{OUT} = 0.1 V or V_{CC} − 0.1 V
		4.5	2.25	1.35 — 1.35		
		5.5	2.75	1.65 — 1.65		
V_{OH}	Minimum High Level Output Voltage	3.0	2.99	2.9 — 2.9	V	I_{OUT} = −50 µA
		4.5	4.49	4.4 — 4.4		
		5.5	5.49	5.4 — 5.4		
		3.0		2.56 — 2.46	V	*V_{IN} = V_{IL} or V_{IH} −12 mA
		4.5		3.86 — 3.76		I_{CH} −24 mA
		5.5		4.86 — 4.76		−24 mA
V_{OL}	Maximum Low Level Output Voltage	3.0	0.002	0.1 — 0.1	V	I_{OUT} = 50 µA
		4.5	0.001	0.1 — 0.1		
		5.5	0.001	0.1 — 0.1		
		3.0		0.36 — 0.44	V	*V_{IN} = V_{IL} or V_{IH} 12 mA
		4.5		0.36 — 0.44		I_{OL} 24 mA
		5.5		0.36 — 0.44		24 mA
I_{IN}	Maximum Input Leakage Current	5.5		±0.1 — ±1.0	µA	V_I = V_{CC}, GND
I_{OLD}	†Minimum Dynamic Output Current	5.5		75	mA	V_{OLD} = 1.65 V Max
I_{OHD}		5.5		−75	mA	V_{OHD} = 3.85 V Min
I_{CC}	Maximum Quiescent Supply Current	5.5		4.0 — 40	µA	V_{IN} = V_{CC} or GND

* All outputs loaded; thresholds on input associated with output under test.
† Maximum test duration 2.0 ms, one output loaded at a time.
Note: I_{IN} and I_{CC} @ 3.0 V are guaranteed to be less than or equal to the respective limit @ 5.5 V V_{CC}.

MC74AC04 • MC74ACT04

AC CHARACTERISTICS (For Figures and Waveforms — See Section 3)

Symbol	Parameter	V_{CC}* (V)	74AC $T_A = +25$ C $C_L = 50$ pF			74AC $T_A = -40$ C to +85 C $C_L = 50$ pF		Unit	Fig. No.
			Min	Typ	Max	Min	Max		
t_{PLH}	Propagation Delay	3.3	1.5	4.5	9.0	1.0	10	ns	3-5
		5.0	1.5	4.0	7.0	1.0	7.5		
t_{PHL}	Propagation Delay	3.3	1.5	4.5	8.5	1.0	9.5	ns	3-5
		5.0	1.5	3.5	6.5	1.0	7.0		

* Voltage Range 3.3 V is 3.3 V ±0.3 V.
  Voltage Range 5.0 V is 5.0 V ±0.5 V.

DC CHARACTERISTICS

Symbol	Parameter	V_{CC} (V)	74ACT $T_A = +25$ C	74ACT $T_A = -40$ C to +85 C	Unit	Conditions
			Typ	Guaranteed Limits		
V_{IH}	Minimum High Level Input Voltage	4.5	1.5	2.0	2.0	V	$V_{OUT} = 0.1$ V or $V_{CC} - 0.1$ V
		5.5	1.5	2.0	2.0		
V_{IL}	Maximum Low Level Input Voltage	4.5	1.5	0.8	0.8	V	$V_{OUT} = 0.1$ V or $V_{CC} - 0.1$ V
		5.5	1.5	0.8	0.8		
V_{OH}	Minimum High Level Output Voltage	4.5	4.49	4.4	4.4	V	$I_{OUT} = -50$ μA
		5.5	5.49	5.4	5.4		
		4.5		3.86	3.76	V	*$V_{IN} = V_{IL}$ or V_{IH} I_{OH} $\quad -24$ mA
		5.5		4.86	4.76		$\quad -24$ mA
V_{OL}	Maximum Low Level Output Voltage	4.5	0.001	0.1	0.1	V	$I_{OUT} = 50$ μA
		5.5	0.001	0.1	0.1		
		4.5		0.36	0.44	V	*$V_{IN} = V_{IL}$ or V_{IH} I_{OL} $\quad 24$ mA
		5.5		0.36	0.44		$\quad 24$ mA
I_{IN}	Maximum Input Leakage Current	5.5		±0.1	±1.0	μA	$V_I = V_{CC}$, GND
ΔI_{CCT}	Additional Max. I_{CC}/Input	5.5	0.6		1.5	mA	$V_I = V_{CC} - 2.1$ V
I_{OLD}	†Minimum Dynamic Output Current	5.5			75	mA	$V_{OLD} = 1.65$ V Max
I_{OHD}		5.5			−75	mA	$V_{OHD} = 3.85$ V Min
I_{CC}	Maximum Quiescent Supply Current	5.5		4.0	40	μA	$V_{IN} = V_{CC}$ or GND

* All outputs loaded; thresholds on input associated with output under test.
† Maximum test duration 2.0 ms, one output loaded at a time.

MC10H103

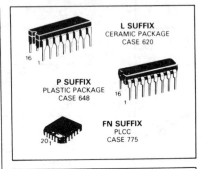

L SUFFIX
CERAMIC PACKAGE
CASE 620

P SUFFIX
PLASTIC PACKAGE
CASE 648

FN SUFFIX
PLCC
CASE 775

QUAD 2-INPUT OR GATE

The MC10H103 is a quad 2-input OR gate. The MC10H103 provides one gate with OR/NOR outputs. This MECL 10KH part is a functional/pinout duplication of the standard MECL 10K family part, with 100% improvement in propagation delay, and no increases in power-supply current.

- Propagation Delay, 1.0 ns Typical
- Power Dissipation 25 mW/Gate (same as MECL 10K)
- Improved Noise Margin 150 mV (Over Operating Voltage and Temperature Range)
- Voltage Compensated
- MECL 10K-Compatible

LOGIC DIAGRAM

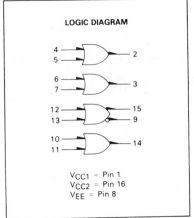

V_{CC1} = Pin 1
V_{CC2} = Pin 16
V_{EE} = Pin 8

MAXIMUM RATINGS

Characteristic	Symbol	Rating	Unit
Power Supply (V_{CC} = 0)	V_{EE}	-8.0 to 0	Vdc
Input Voltage (V_{CC} = 0)	V_I	0 to V_{EE}	Vdc
Output Current — Continuous — Surge	I_{out}	50 100	mA
Operating Temperature Range	T_A	0–+75	°C
Storage Temperature Range — Plastic — Ceramic	T_{stg}	-55 to +150 -55 to +165	°C °C

ELECTRICAL CHARACTERISTICS (V_{EE} = -5.2 V ±5%) (See Note)

Characteristic	Symbol	0° Min	0° Max	25° Min	25° Max	75° Min	75° Max	Unit
Power Supply Current	I_E	—	29	—	26	—	29	mA
Input Current High	I_{inH}	—	425	—	265	—	.265	μA
Input Current Low	I_{inL}	0.5	—	0.5	—	0.3	—	μA
High Output Voltage	V_{OH}	-1.02	-0.84	-0.98	-0.81	-0.92	-0.735	Vdc
Low Output Voltage	V_{OL}	-1.95	-1.63	-1.95	-1.63	-1.95	-1.60	Vdc
High Input Voltage	V_{IH}	-1.17	-0.84	-1.13	-0.81	-1.07	-0.735	Vdc
Low Input Voltage	V_{IL}	-1.95	-1.48	-1.95	-1.48	-1.95	-1.45	Vdc

AC PARAMETERS

		0° Min	0° Max	25° Min	25° Max	75° Min	75° Max	Unit
Propagation Delay	t_{pd}	0.4	1.3	0.4	1.3	0.45	1.45	ns
Rise Time	t_r	0.5	1.7	0.5	1.8	0.5	1.9	ns
Fall Time	t_f	0.5	1.7	0.5	1.8	0.5	1.9	ns

NOTE:
Each MECL 10 KH series circuit has been designed to meet the dc specifications shown in the test table, after thermal equilibrium has been established. The circuit is in a test socket or mounted on a printed circuit board and transverse air flow greater than 500 lfpm is maintained. Outputs are terminated through a 50 ohm resistor to -2.0 volts.

DIP PIN ASSIGNMENT

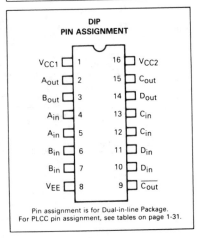

Pin assignment is for Dual-in-line Package.
For PLCC pin assignment, see tables on page 1-31.

3.6 Concluding comments and further reading

This chapter has described what is inside the integrated circuits used in the major logic technologies. As a designer you will often find it necessary to be aware of the properties of the circuits you are using to implement designs. You will be designing circuits to meet a wide-ranging specification. Some aspects of the specification are often more critical than others, and these can dominate the logic technology chosen for implementation. Designers use fewer committed circuits now (those available from logic families), but you will doubtless resort to them from time to time. Though most designers use programmable logic implementations, these naturally depend on a logic technology for implementation. There are other logic technologies; some are in a development stage, some are now obsolete, while others have not penetrated the market much. Failure to penetrate the market may imply that the technology did not live up to its early expectations, but there are also commercial considerations that are beyond the scope of this text.

For an introduction to semiconductor physics, try Parker, G. J., *Introductory Semiconductor Device Physics* (Prentice-Hall, 1993) or Millman, J. and Grabel A., *Microelectronics* (McGraw-Hill, 2nd edn, 1988) which gives a lucid account of microelectronic technology. There are a number of reasonably priced texts which include digital electronic circuits within a complete study of electronics: see, for example, Sedra, A. and Smith, K., *Microelectronic Circuits* (Saunders, 1991). Haznedar, H., *Digital Microelectronics* (Benjamin, 1991) gives an up-to-date and detailed presentation of logic technologies. There are books dedicated to particular technologies: see, for example, Elmasry, M. I., *Digital Bipolar Integrated Circuits* (Wiley, 1983), Uyemura, J. P., *Fundamentals of MOS Integrated Circuits* (Addison-Wesley, 1988), Weste, N. H. E. and Estraghian, K., *Principles of CMOS VLSI Design* (Addison-Wesley, 2nd edn, 1992) and Alvarez, A. R., *BICMOS Technology and Applications* (Kluwer Academic Publishers, 1989).

Manufacturers provide databooks on the circuits that they produce, as well as handbooks on how to use them. Arguably the most famous logic databook is the *TTL Handbook* (Texas Instruments Inc., 4 vols), but there are handbooks (for example, those by Motorola Inc. and Philips) for Bipolar, CMOS and ECL logic families. There are many manufacturers, and many vendors (e.g. RS Components and Farnell Instruments) will provide datasheets on products that they supply.

3.7 Questions

1 Develop a CMOS circuit to implement a 2-input (positive logic) AND function.

2 Provide a TTL circuit for a 2-input (positive logic) AND function.

3 How is a 2-input ECL AND gate achieved?

4 A DTL NAND gate is made from a diode logic AND gate coupled to an inverting transistor amplifier by two coupling diodes, rather than a single coupling diode. For the case where all the diodes' cut-in voltage is 0.5 V and for the transistor $V_{BE}(ON) = 0.6$ V, explain any effect the extra coupling diode has on the noise margins N_{MH} and N_{ML}.

5 An NMOS NAND gate is made from two transistors: one with ON resistance of 200 ohms and the other with an ON resistance which can be controlled. The OFF resistance for both transistors is in the order of $G\Omega$. What ratio of ON resistance must be achieved so that a LOW output approaches $V_{DD}/3$?

6 Design a CMOS (positive logic) circuit to implement the function

$$f = \overline{A \cdot B + C \cdot D}$$

7 An ECL gate has a positive and negative voltage supply of $+1.0$ V and -4.0 V, respectively.

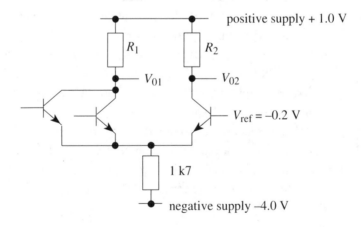

The input voltage levels are $V_{IH} = 0.2$ V for a '1' and $V_{IL} = -0.6$ V for a '0'. Given that the transistors switch ON for $V_{BE}(ON) = 0.8$ V and that the voltage difference between a '1' and a '0' is the same at either output as it is at either input, calculate the values of the resistors R_1 and R_2. Show also that no transistor enters saturation.

4   Introductory Sequential Logic

'This is not the end. It is not even the beginning of the end. But it is, perhaps, the end of the beginning.'

Winston Churchill

4.1   Sequential logic concepts

We will now consider systems of logic that act in *sequence*. These systems find wide application, such as, for example, in implementing a sequence of instructions in a computer, or in implementing systems with a sequence of events such as an industrial product control system. This implies that:

(a) we have a synchronising signal, called a *clock*, which is similar to a clock in everyday life;
(b) we have to remember where we are in the sequence; this requires *memory*.

Sequential logic then concerns binary systems that go through a sequence of *states* where the state is the condition at a particular time. Progression between states is synchronised to a clock signal. The clock signal indicates when to move from one state to the next and is usually a square-wave signal:

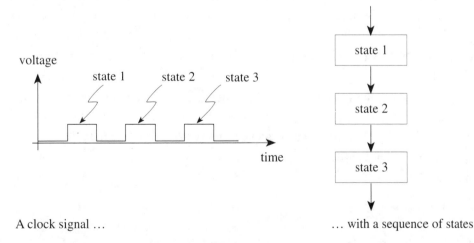

A clock signal ...

... with a sequence of states

85

In practice there are usually more than three states, with a more complex routing structure. The states change when the clock is '1'. This is *level-triggered* logic, since we are synchronising to a **level** of the clock. When the clock is HIGH we **change** state; when the clock is LOW we **remember** the state. We now have a sequencer, the clock, and devices that have **memory**. These devices are called *bistables*.

4.2 Bistables

A bistable is defined as a device with two stable states (as implied in its name). In its basic form it comprises a pair of inverters (labelling the outputs Q and $\overline{Q}$ is a convention for bistables since they are usually complementary and one is the inverse of the other):

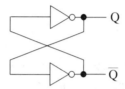

This bistable is perfectly stable in either of two states:

(a) when Q = '1' and $\overline{Q}$ = '0'
(b) when Q = '0' and $\overline{Q}$ = '1'

It can enter either of these states when power is applied. The state in which it actually ends up depends on the propagation delays of the inverters (whichever of the inverters responds fastest determines the final state of the bistable). The final state is actually controlled by *feedback*, since the output of each inverter forms the input to the other. This feedback provides the memory associated with the device. The two memory states are called *set* and *reset*. A bistable is **set** when Q = '1' and $\overline{Q}$ = '0' and **reset** when Q = '0' and $\overline{Q}$ = '1'.

Since we cannot control the memory of this device, we replace the inverters with gates that accept more than one input. We shall replace the inverters with NAND gates and retain the feedback or *steering network*, since this provides the memory by feeding back the current state to the inputs.

4.2.1 $\overline{R}\overline{S}$ *bistable*

The $\overline{R}\overline{S}$ *bistable* is formed from two cross-coupled NAND gates:

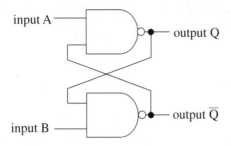

input A — output Q

input B — output $\overline{Q}$

This looks like a combinational logic system, but let us consider what happens when the input A is a '0' and is then changed to a '1' while input B remains unchanged at '1'. These are labelled as input A = 0/1 ('0' changing to a '1') and input B = 1/1 ('1' remaining at '1'). Noting that the output of a NAND gate is '1' if any input is a '0', the initial state of the bistable will be Q = '1' (since input A = '0') and $\overline{Q}$ = '0' (since $\overline{Q}$ is the output of a NAND gate with inputs Q = '1' and B = '1'). When input A changes from '0' to '1' then Q = '1' since $\overline{Q}$ = '0'; also, $\overline{Q}$ = '0' since Q = '1'. The bistable does not change state (the outputs remain unaffected by the change in input A), and can be interpreted to have memory since the bistable remembers that it was set. This memory is provided by the steering network feeding the outputs back to become inputs, which makes $\overline{Q}$ force Q to remain at '1'.

Similarly, if input A is held at '1' and input B changes from a '0' to '1', then the outputs will initially be $\overline{Q}$ = '1' (since input B = '0') and Q = '0'. The outputs again remain constant and do not change when B changes since Q = '0' is fed back to keep $\overline{Q}$ = '1'. The memory function associated with changing both inputs HIGH can be summarised as follows:

Inputs		Outputs	
A	B	Q	$\overline{Q}$
0/1	1/1	1/1	0/0
1/1	0/1	0/0	1/1

This is termed *latching* – the $\overline{RS}$ bistable latches, or remembers, the outputs when both inputs are HIGH. Bistables usually have **two** states:

(a) **set** when Q = '1', $\overline{Q}$ = '0'
(b) **reset** when Q = '0', $\overline{Q}$ = '1'

so this bistable latches whether it was set or reset. Input A = '0' sets the device, since it forces the output Q to be '1'; it is then usually called $\overline{S}$, since a LOW value sets the device. Input B = '0' resets the device, since it forces $\overline{Q}$ to be '1' (and Q to be '0'); it is then usually called $\overline{R}$, since a LOW value resets the device.

Note that you cannot form memory elements without inversion, and they cannot be made from AND or OR gates alone.

The operation of the bistable is described by a *characteristic table*, which defines the new state of the bistable Q_{n+1} and $\overline{Q}_{n+1}$ as either set, reset or latched (when the new state of the bistable Q_{n+1} latches the previous state Q_n, $Q_{n+1} = Q_n$ and $\overline{Q}_{n+1} = \overline{Q}_n$):

$\overline{S}$	$\overline{R}$	Q_{n+1}	$\overline{Q}_{n+1}$	
0	0	1	1	Invalid
0	1	1	0	Set
1	0	0	1	Reset
1	1	Q_n	$\overline{Q}_n$	Latch

When $\overline{S} = \overline{R} = 0$ the outputs are *invalid*, since they are both '1' and the bistable appears to be set and reset at the same time. This is sometimes called **unstable** (it is manifestly not unstable, since both outputs are fixed HIGH) since if $\overline{S} = \overline{R} = 0$ and we then change to $\overline{S} = \overline{R} = 1$ then the output state will depend on propagation delays within the device (akin to a basic bistable formed from inverters). For $\overline{S}, \overline{R} = 0,1$ and $\overline{S}, \overline{R} = 1,0$ the bistable is set and reset respectively. In this state the device acts as a combinational logic system; the bistable only exhibits memory when $\overline{S}, \overline{R} = 1,1$.

An alternative formulation of the $\overline{RS}$ bistable is to replace NAND with NOR, giving an RS bistable:

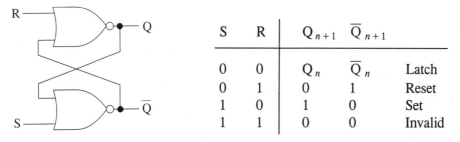

S	R	Q_{n+1}	$\overline{Q}_{n+1}$	
0	0	Q_n	$\overline{Q}_n$	Latch
0	1	0	1	Reset
1	0	1	0	Set
1	1	0	0	Invalid

The set and reset states are entered when S and R are '1' respectively. The outputs latch for S = R = '0' and are invalid when both inputs are HIGH.

4.2.2 Switch debouncer – an $\overline{RS}$ bistable in action

This is a commonly cited application of a basic bistable. It centres on using a bistable to remove the effects of bouncing in a mechanical switch. Ideally, when a mechanical switch changes from $+5\,V$ to $0\,V$ the voltage should change with it:

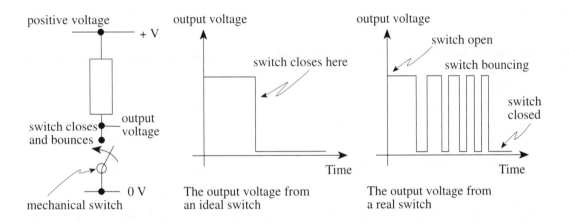

Unfortunately, mechanical switches bounce and the voltage will change from $+5\,V$ to $0\,V$ and back again several times. This happens on all mechanical switches, even domestic light switches, but these effects are so fast that they are imperceptible to the human eye. They can, however, have unfortunate consequences when used in sequential logic systems.

The $\overline{RS}$ bistable can be used to debounce a single-pole, double-throw switch, to remove the extra transitions in the output voltage resulting from mechanical bounce. The circuit is as follows:

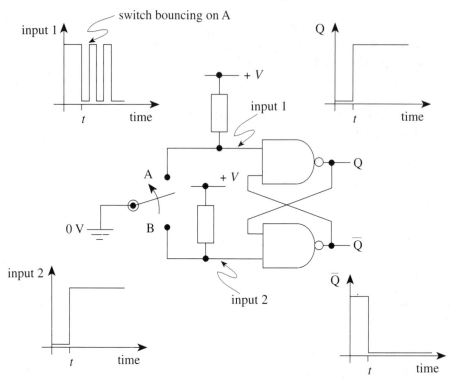

When the switch is thrown from B to A it bounces on A. This means that input 2 is a '1', since B is unconnected and there is a *pull-up* resistor pulling input 2 to HIGH. Conversely, input 1 is changing between a '1' and a '0' (when the switch connected it is '0' and when the switch bounces up it is '1'). However, the memory function of bistables (via the **pull-up** resistors) ensures that outputs change only once when the switch position is changed. The bistable outputs track the first change in the switch position, and when the switch bounces this state is latched.

> *This type of switch is known as a single-pole, double-throw switch. The single pole is the earth connection and the (double) throw is the switch connections to A and B. Push-button switches (single pole, single way, momentary action), as used in the logic circuits earlier, require a different debouncing circuit since there is no common pole for reference for the bistable.*

4.2.3 D-type bistable

Returning to the basic $\overline{RS}$ bistable, the main difficulty is the invalid state where both outputs are the same (the device is set and reset at the same time). One solution is to use a single input and invert it between the $\overline{R}$ and $\overline{S}$ inputs. This is the D-type bistable:

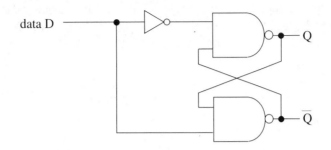

Its function is again summarised by a characteristic table:

D	Q	$\overline{Q}$
0	0	1
1	1	0

This shows that we have avoided the invalid state. However, the basic D-type is functionally useless, since it merely **delays** the input and an equivalent circuit is

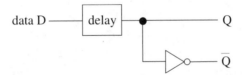

4.2.4   Level-triggered D-type

The basic D-type becomes useful when you add in a clock to give you a *level-triggered D-type*. The function of this bistable is to accept data when the clock is *asserted* at '1' and to latch the data when the clock is *disasserted*. The clock is conventionally gated using NAND gates to allow the data input to affect the bistable output only when the clock is '1'. These NAND gates are coupled to an $\overline{RS}$ bistable as

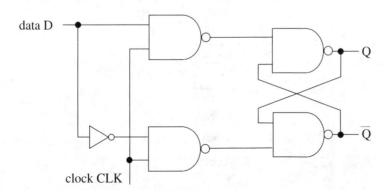

When CLK = '0' we remember the previous value of the outputs, since with the clock at '0' the outputs of both NAND gates are '1' and we are then using the latch state of the $\overline{RS}$ bistable. When CLK = '1' the input can change the outputs:

D	CLK	Q_{n+1}	$\overline{Q}_{n+1}$
X	0	Q_n	$\overline{Q}_n$
0	1	0	1
1	1	1	0

We now have a *1-bit memory* (it latches one bit when the clock is asserted). In order to remember the bit of information, we must first set up the data (the bit we want to remember) then clock the device. The bit will then be latched until the following clock cycle. Since the outputs change with the inputs when the clock is asserted (HIGH) it is also called a *transparent D-type latch*.

4.2.5   Level-triggered RS bistable (transparent RS latch)

A clock input can similarly be connected into an $\overline{RS}$ bistable to give a *level-triggered RS bistable*.

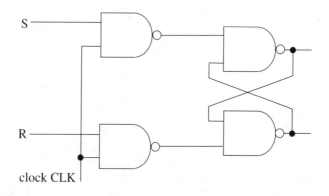

Its characteristic table is:

S	R	CLK	Q_{n+1}	$\overline{Q}_{n+1}$	
X	X	0	Q_n	$\overline{Q}_n$	Latch
0	0	1	Q_n	$\overline{Q}_n$	Latch
1	0	1	1	0	Set
0	1	1	0	1	Reset
1	1	1	1	1	Invalid

This provides the functionality of the $\overline{RS}$ bistable only when the clock is asserted. As with the level-triggered D-type the outputs will change with any change in the inputs for CLK . H and so it is also a **transparent latch**. Note also that S = '1' and R = '0' sets the bistable (for a clock = '1'), R = '1' and S = '0' resets it. Note also there is still an **invalid** state (for R = S = '1', Q = Q = '1').

4.2.6   *Level-triggered JK bistable*

The *JK bistable* avoids the invalid state by feeding the outputs back twice, i.e. using a double steering network. The invalid state is replaced with a *toggle* state where the outputs flip to become the complement of each other, $Q_{n+1} = \overline{Q}_n$, $\overline{Q}_{n+1} = Q$. All other states are the same as for the $\overline{RS}$ type.

> *The D in the D-type stands for delay, the $\overline{RS}$ bistable is well labelled, but what about the JK? Who knows – the origins of the JK appellation are forgotten now.*

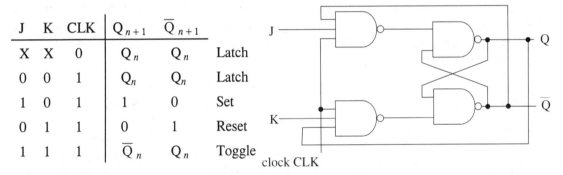

J	K	CLK	Q_{n+1}	$\overline{Q}_{n+1}$	
X	X	0	Q_n	Q_n	Latch
0	0	1	Q_n	Q_n	Latch
1	0	1	1	0	Set
0	1	1	0	1	Reset
1	1	1	$\overline{Q}_n$	Q_n	Toggle

clock CLK

If you analyse this circuit you will find that it actually toggles continually when both inputs are HIGH and the clock is asserted. It then has no function as a level-triggered device but is included (again) for completeness within the set of basic bistables. The JK bistable is rather archaic now, and design centres mostly on use of D-type bistables. If you need a JK, it is easy to make one from a D-type, and this will be illustrated in Section 5.6.1.

4.3   Clocks and synchronism

We have so far considered level triggering where events are synchronised to a HIGH level of the clock. When the clock is LOW, the device is latched and the outputs are static and do not change with the inputs. Synchronism could easily be based on a LOW active clock level with devices latched for a HIGH clock. It does not really matter which way round the clock is. A major problem is that the outputs can change all the time the clock is asserted.

We usually need a more precise time for synchronism, to synchronise to a much smaller portion of the clock. For this reason we use the edge of the clock; this technique is termed *edge-triggering*:

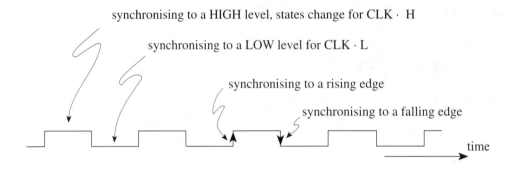

synchronising to a HIGH level, states change for CLK · H

synchronising to a LOW level for CLK · L

synchronising to a rising edge

synchronising to a falling edge

time

There is a range of symbols that indicate whether the part of the clock used for triggering is an active HIGH level (outputs change with inputs for CLK . H), active LOW level (outputs can change for CLK . L), positive edge (the outputs assume the value of the inputs just prior to the clock changing from LOW to HIGH), or negative edge (where outputs assume input values just prior to the falling edge of a clock). *Synchronous sequential systems* are those where all sequential devices are driven by the same clock event. For many systems this concerns edge-triggering, and all elements are triggered at exactly the same time.

> *There are asynchronous circuits too. In these circuits, elements are not all driven by the same clock event. Also, the term bistable is sometimes inferred to mean an edge-triggered bistable, and latch to mean a level-triggered version. The term bistable is now predominantly ubiquitous and refers to either type.*

These symbols are:

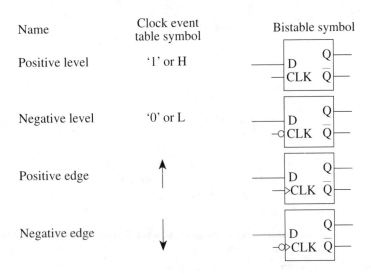

Name	Clock event table symbol	Bistable symbol
Positive level	'1' or H	
Negative level	'0' or L	
Positive edge	↑	
Negative edge	↓	

4.4 Master–slave D-type bistable

The master–slave bistable was one of the earliest forms of edge-triggering. It uses two D-type transparent latches where the clock of one is inverted to provide the clock for the other. The two D-types are connected in series:

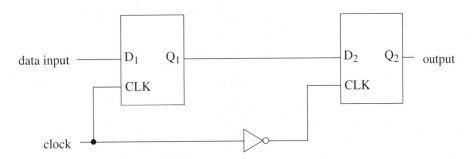

The inversion of the clock between the two stages provides a form of edge-triggering, since the first D-type is enabled when the clock is HIGH and the second D-type afterwards, when the clock is LOW. Then the output of the first bistable will follow any change in the input when the clock is HIGH, but will be fixed to their last state (when the clock was HIGH) when the clock returns LOW and the first bistable latches. This value will be latched by the second bistable, since its own clock is now HIGH. The phases of the clock signal are as follows:

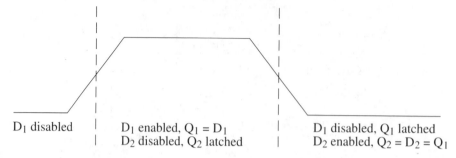

D_1 disabled	D_1 enabled, $Q_1 = D_1$	D_1 disabled, Q_1 latched
	D_2 disabled, Q_2 latched	D_2 enabled, $Q_2 = D_2 = Q_1$

The output Q_1 (and hence the input D_1) **just prior** to the falling clock edge is latched as the output of the device. By inverting the clock between bistables the inputs never directly affect the output, avoiding the **transparency** problem in the basic latches. The outputs of the second bistable are a latched version of the outputs of the first and are constant. Though the device actually latches data values just prior to the falling edge of the clock, the master–slave bistable is not a true edge-triggered device since it needs both halves of the clock; it is actually **pulse-triggered**, though some books refer to it as edge-triggered.

The name 'master–slave' is perhaps the nastiest in digital electronics. Is one the whipping post for the other?

Its characteristic table is

D	CK	Q_{n+1}
X	0	Q_n
X	1	Q_n
1	⎍	1
0	⎍	0

The timing for the master–slave bistable is then

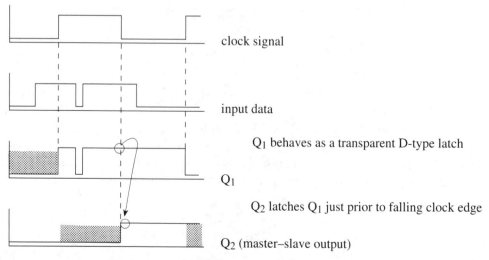

clock signal

input data

Q_1 behaves as a transparent D-type latch

Q_1

Q_2 latches Q_1 just prior to falling clock edge

Q_2 (master–slave output)

Shaded areas indicate where Q_1 and Q_2 are latched.

4.5 Positive edge-triggered D-type

This device is a true edge-triggered device. Input data just prior to a rising clock edge becomes the outputs when the clock changes from '0' to '1'. Input data is then transferred to the outputs at the positive edge of the clock signal. This synchronises change to an extremely small portion of the clock, which has particular advantages in practical designs. Its symbol and operation, summarised using a characteristic table, are:

D	CLK	Q_{n+1}
X	0	Q_n
X	1	Q_n
1	↑	1
0	↑	0

This characteristic table can also be interpreted as follows:

Present state		Next state
Q	D	Q
X	1	1
X	0	0

where the transition between present and next state occurs at **active clock transitions**. This allows us to consider the output of a sequential device to depend on the current value of the output (the current state) and the current value of the inputs. We shall use such concepts later in the analysis and design of sequential systems.

4.6 Timing considerations

All timing is specified **relative to the clock**. The main timing considerations refer mainly to edge-triggered logic. This is because of the transparency problem associated with level-triggered systems. Owing to the problems associated with transparency in large-scale systems, edge-triggering is often preferred and it is then better to treat timing considerations pertinent to edge-triggered logic alone. Clocks do not change infinitely fast, and so timing is usually referred to a **threshold** point and manufacturers often use a 50% point, half-way between HIGH and LOW. Data must be **set up** before the clock edge, and T_{SU} (T_{SET-UP}) defines how long before the clock edge the data must be ready and stable. Data must be **held** constant after the clock edge and T_H (T_{HOLD}) defines for how long data must be held.

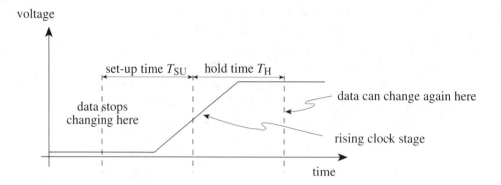

In some devices, changing data inputs just after the timing edge can cause the wrong data to be latched, but in many the hold time, T_H, is zero. In most devices, violating the set-up time, T_{SU}, may cause *unpredictable behaviour*; this is often referred to as *metastability*. The bistable input data should never change within the window defined by the set-up and hold times.

> *Metastability describes a bistable output between its stable states (HIGH or LOW). Any bistable output can reach such a state, and though momentary it can have disastrous consequences.*

Changes in the inputs take some time to propagate through to the outputs, and this gives rise to *propagation delay*, which can differ between the output changing from LOW to HIGH and from HIGH to LOW. The propagation time for an output changing from HIGH to LOW, T_{PHL}, and the propagation delay for an output changing from LOW to HIGH, T_{PLH}, define how long after the clock edge outputs will change, either from HIGH to LOW or from LOW to HIGH respectively. This is effectively how long you have to wait to see a result. These times are usually quoted by manufacturers for devices as **worst-case** to guarantee a product's performance. Together, these timing parameters then give the maximum clock frequency at which you can drive devices.

The sequence of events for a bistable in a sequential system are:

(a) the data is set up;
(b) after the clock event, the data propagates to the outputs;
(c) the outputs propagate through any combinational logic;
(d) the data is set up again ready for the next clock event.

The **minimum clock period** is then the total time for the first three elements of this sequence (the fourth contributes to the following clock cycle). The minimum clock period T_{min} is the reciprocal of the maximum clock frequency, f_{max}, and is given by

$$T_{\min} = \frac{1}{f_{\max}} = T_P + T_C + T_{SU}$$

where T_P is the larger value of T_{PLH} or T_{PHL}. This is the worst-case value, the longest time needed to respond in calculating the **maximum possible frequency** at which we can issue clock edges and still obtain a satisfactory output. The combinational logic delay, T_C, implies that the next state of the bistable depends on a combination of logic signals; again this should be a worst-case value – the longest path that any combinational signals affecting the input have to traverse, following a clock edge. Since the hold time is invariably less than the propagation time, it does not affect the minimum clock period and therefore does not appear it its calculation.

The following circuit will toggle when the enable signal is HIGH and be continuously LOW when enable is LOW:

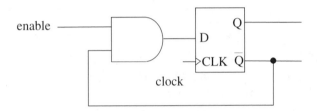

The maximum clock frequency for this circuit can be calculated from the timing parameters. For the bistable $T_{PLH} = 10$ ns, $T_{PHL} = 12$ ns and $T_{SU} = 3$ ns, and the worst-case propagation delay of the AND gate, T_P (AND), is 10 ns. The minimum clock period is then

$$T_{\min} = T_{PHL} + T_P \text{(AND)} + T_{SU} = 25 \text{ ns}$$

This corresponds to a maximum clock frequency of 40 MHz.

4.7 Asynchronous inputs

When you first apply power to a bistable it will be either set or reset according to propagation delays in the device; most bistables do not conveniently reset when you apply power. Two *asynchronous inputs* can be used to control the state of a bistable without reference to the clock signal (this is why they are called asynchronous, since their action is not synchronised with the clock):

preset: $\overline{PR}$ = '0' (active LOW) sets the bistable and sets Q HIGH
clear: $\overline{CLR}$ = '0' clears it and resets Q LOW

The last stage of a bistable is an $\overline{RS}$ bistable and this is where the asynchronous inputs are included:

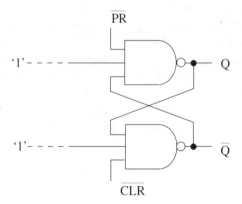

When bistable outputs are stable (after the propagation time is completed) the last $\overline{RS}$ stage is then latched. The preset input, $\overline{PR}$, can be used to set the bistable; the clear input resets it. These inputs should always be complementary when used, and their function is usually included within a characteristic table as

Asynchronous inputs		Clock	Output
$\overline{PR}$	$\overline{CLR}$	CLK	Q
0	1	X	1
1	0	X	0

These inputs are **asynchronous**, not linked to the clock, and should be used with **caution**. Note that asynchronous inputs can be **synchronised** using a **latch**. If they are not used it is best to connect them to their inactive state.

4.8　Sequential systems

4.8.1　*General sequential system*

The general form of a sequential system is one which moves through a sequence of states, entering each new state at active clock events. The next state following a clock event is determined by the current position in the sequence and by external inputs. Each state is indicated by the bistables within a system, and the bistables will be either set or reset in a particular state. The current state is latched by the bistables and the bistable outputs give the current state. In order for this to be used to determine the next state, the outputs must be fed back as bistable inputs. These are combined with any external inputs to determine the next state of the system. The general schema is then

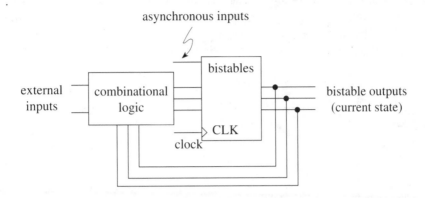

The combinational circuitry, combining the current state or bistable outputs with external or control inputs, is often referred to as the *next-state logic*. There are *formal techniques* to design the next-state logic. The formal techniques will be described after we have dealt with sequential system software specification, and made an introductory study of basic sequential devices to highlight how they operate.

> *There are a number of versions of this diagram. The inputs can be combined with gated outputs or by using multiplexers. All versions achieve the same aim.*

4.8.2 Sequential system software specification

In order to provide a software specification of a sequential system we need to introduce triggering or synchronism into the software. The design of most large-scale sequential circuitry now uses edge-triggered D-type bistables for most implementations. For D-types, data is transferred to the outputs from a single input. The data is therefore **assigned** to the output and so for sequential system software specification we shall use the assignment operator, $=$, to denote the clock event when assigning values to bistable, termed *registered* (as explained later in Section 4.8.3) outputs. The software description for an edge-triggered D-type is then

```
IF (data) THEN Q_output=HIGH ELSE Q_output=LOW
```

however, since the registered output follows the input directly it is easier to describe a D-type as merely

```
output = data
```

and the output assumes the data input at clock events. In a sequential system the next state could be a function of the current state only, without external inputs. In such systems, the current inputs to the bistables are a fedback combination of their outputs. An appropriate software description is to use a CASE statement, where the next state is determined from a look-up table of current states. The look-up table comprises the condition of each bistable, whether it is set or reset. To implement the sequence of states

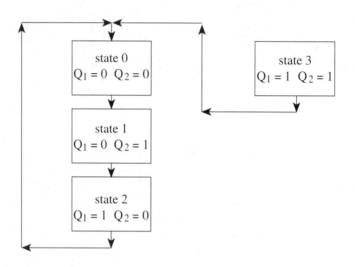

the software description could be

```
CASE (current state)
BEGIN
  #b00: next_state=state_1
      ;if q1 and q2 are both 0 then go to state 1 at next clock edge
  #b01: next_state=state_2
      ;if q1=0 and q2=1 then go to state 2 at next clock edge
  #b10: next_state=state_0
      ;if q1=1 and q2=0 then go to state 0 at next clock event
  #b11: next_state=state_0
      ;if q1=q2=1 then go to state 0 at next active clock
END
```

In this description #b indicates testing the binary value of the two bistable outputs, though it is not yet explicit as to which is tested. It is also unclear as to how the current state refers to the bistable values. We need to describe bistable outputs using a *vector*, which gives a shorthand way of referring to all bistable values, i.e. to the current state of the device. Using a vector of two bistable outputs requires combining them within the vector. For two outputs Q_1 and Q_2 we could combine them in a vector called outputs, which is of dimension 2 (has two elements). We can then use the first element outputs[1] to refer to Q_1, and the second element outputs[2] to refer to Q_2. When referring to the state of the device we can refer to outputs[1..2], and this can be used either to return the current state of the device (whether the bistables are set or reset) or else to assign a new state. The numbers after the #b then refer to Q_1 first and Q_2 second. The software description for this system then becomes

```
CASE (outputs[1..2])
BEGIN
    #b00: outputs[1..2]=#b01
        ;if q1 reset and q2 reset then at next clock edge q2 is set
    #b01: outputs[1..2]=#b10
        ;if q1 reset and q2 set then set q1 and reset q2
    #b10: outputs[1..2]=#b00
        ;if q1 set and q2 reset then reset q1 at next clock event
    #b11: outputs[1..2]=#b00
        ;if q1 and q2 both set then reset both at next active clock
END
```

If external inputs are included, we need to decide what action to take. We then need to include a decision box within the chart describing the sequence of states and we choose an appropriate branch in the chart according to the value of the tested variable or input. The following chart shows a system with a different sequence of states for different values of the external input. At each decision box the external control input is tested and, if it is HIGH, then the system moves to a state different from that entered when the control input is LOW.

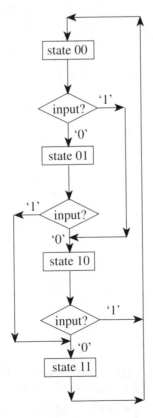

The software description is then

```
CASE (outputs[1..2])
BEGIN ;this description is derived from the state chart
  #b00: IF (input) THEN outputs[1..2]=#b10
        ELSE outputs[1..2]=#b01
  #b01: IF (input) THEN outputs[1..2]=#b11
        ELSE outputs[1..2]=#b10
  #b10: IF (input) THEN outputs[1..2]=#b00
        ELSE outputs[1..2]=#b11
  #b11: outputs[1..2]=#b00
  ;always returns to state 00 after state 11
END
```

> *The IF statement could have been written outside the CASE statement to provide a different software description but one which describes the same function.*

4.8.3   Registers

A *register* is merely a set of latches connected in parallel, all driven by the same clock. Naturally, its function is to **register** or latch data presented in **parallel**. In computer architectures information is stored as a pattern of bits. This information can be the program that the computer is designed to execute or the data that it is intended to process. We need to latch data as part of a processing cycle, and this is achieved using a register. It is then a basic part of any sequential system:

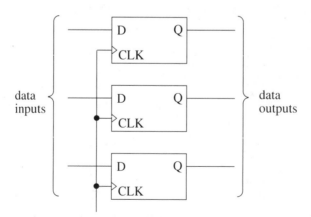

This can use either level-triggered or edge-triggered bistable elements. There is no combinational circuitry associated with this register; in operation we first set up the data, then issue a clock event to transfer the data to the contents of the register, and then hold the data for some time after (according to the timing parameters associated with the device). The register then contains the information that we designed it to hold. Since it stores a number of bits all presented at the same time, it then has the facility to register *parallel data*.

This is why bistable outputs are referred to as registered outputs in software. For a vector of registered outputs of D-type bistables, `outputs[1..3]`, and a vector of inputs, `inputs[1..3]`, a software description assigns data to become outputs following a clock edge and is then simply

```
outputs[1..3]=inputs[1..3]  ;register where each output takes value
                            ;of associated input
```

4.8.4 Shift register

A *shift register* is a device which allows you to move (shift) data within it. The data is shifted from one bistable to the next and its architecture is then a series of bistables with outputs connected to become the input to the next bistable:

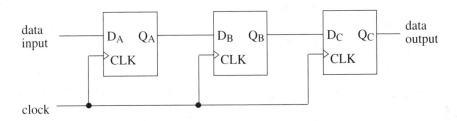

Assuming that all bistables are cleared before we start (using the asynchronous inputs), then given data which is '0' at the first rising clock edge, '1' at the next, '0' at the third and '1' at the fourth and fifth clock edges, and '0' thereafter, we can depict the data as a time-varying signal as follows:

This is synchronised to a clock signal:

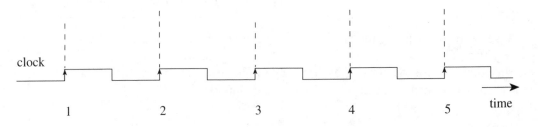

The arrows on the clock signal signify that the timing charts refer to a positive edge-triggered system. For analysis we shall assume that all bistables are reset with Q = '0' prior to the first clock pulse. At the first clock pulse the input to bistable A will be '0' (the input data) and the inputs to all other bistables will be '0', since all bistables are reset. At the first clock edge no bistable will change state, since no input changes and all bistables will remain reset. At the second clock pulse the input to the bistable A will be '1' (the input data), whereas the inputs to all others will still be '0'. After the second clock pulse bistable A output Q_A will be '1' but all others will remain at '0'. At the third clock pulse the input to bistable A will be '0' and the input to bistable B will be '1', since Q_A = '1'. All other bistable inputs will still be '0'. After the third clock pulse, Q_A = '0' and Q_B = '1', and all other bistable outputs are '0'. The '1' remembered by bistable A at the second clock pulse has then moved, or shifted to, bistable B at the third clock pulse. Input data then moves from left to right by one clock pulse at each clock event. This can be summarised in a timing diagram as follows:

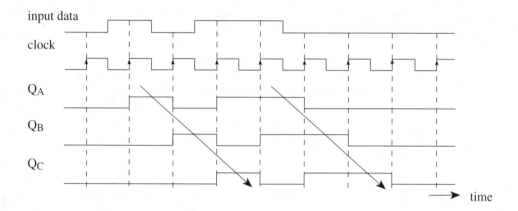

Note that the shift register accepts input data one bit at a time and it is then said to latch *serial data*. The initial diagram and subsequent analysis has been based on edge-triggered elements. Were you to attempt to build a shift register from level-triggered elements, then problems associated with transparency would arise. Since the outputs of a transparent latch can change whenever the clock is asserted, the first input data '1' will appear at the output Q_A and go immediately to the input of bistable B, appear at Q_B and pass to Q_C. If level-triggered bistables were used, data would simply fill up the entire register and no functionality would ensue. A shift register **must** use edge-triggered bistable elements.

A shift register implements the shift instructions used in assembly language programming.

A software description for a vector of registered outputs, outputs[1..3], where outputs[1] = Q_A, outputs[2] = Q_B, and outputs[3] = Q_C. The software description can use the state of each bistable (whether it is SET or RESET), noting that the registered output outputs[1] follows the input data:

```
;3-bit shift register
CASE (outputs[1..3])
BEGIN  ;in the next state:q1=data, q2=q1, q3=q2
#b000: IF (data)  THEN outputs[1..3]=#b100 ;here the data = 1
                  ELSE outputs[1..3]=#b000 ;data = 0
#b001: IF (data)  THEN outputs[1..3]=#b100
                  ELSE outputs[1..3]=#b000
#b010: IF (data)  THEN outputs[1..3]=#b101
                  ELSE outputs[1..3]=#b001
#b011: IF (data)  THEN outputs[1..3]=#b101
                  ELSE outputs[1..3]=#b001
#b100: IF (data)  THEN outputs[1..3]=#b110
                  ELSE outputs[1..3]=#b010
#b101: IF (data)  THEN outputs[1..3]=#b110
                  ELSE outputs[1..3]=#b010
#b110: IF (data)  THEN outputs[1..3]=#b111
                  ELSE outputs[1..3]=#b011
#b111: IF (data)  THEN outputs[1..3]=#b111
                  ELSE outputs[1..3]=#b011
END
```

An alternative software description could use the functionality of the device and is much simpler. Given the same vector of outputs, the description could be

```
;3-bit shift register
BEGIN
outputs[1]=data
outputs[2]=outputs[1]
outputs[3]=outputs[2]
END
```

This is clearly a much simpler description, but is achieved only by intimate knowledge of the functionality of the shift register and as such can only be achieved for simple systems.

Bidirectional shift register

The shift register just presented can only shift data from left to right. This is clearly schematic according to the book layout, and we could equally have a shift register which shifts the other way:

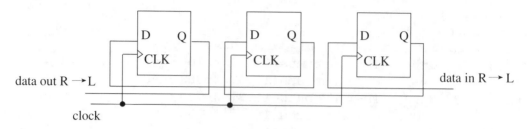

It is often required to have a shift register which shifts information either way, either from left to right or from right to left, according to the status of a mode signal. We then need to choose as a bistable input either data coming from the left or data from the right and a circuit to ensure that the appropriate data is selected. We need the function of a multiplexer, the ability to select either of two inputs according to a mode signal. By attaching a multiplexer to each input and by driving the multiplexers with a common control signal we then achieve a register with bidirectional functionality, which shifts data right to left if the mode signal $\overline{L}/R$ is HIGH and left to right when $\overline{L}/R$ is LOW:

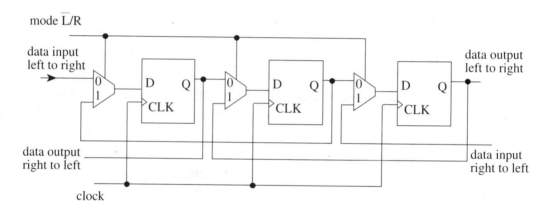

This sequential system now has a state that corresponds to the data stored within the bistables. The next state depends on the current state, together with the status of an (external) mode signal. We are then combining bistable outputs with external inputs to determine the sequence of states of the bistable. This then accords with the general schema of a sequential system described earlier. Sequential systems rarely fit into a prescribed register pattern, so it is now appropriate to develop a formal technique for sequential system design and synthesis.

4.9   Algorithmic state machine (ASM) design

4.9.1   The ASM method

In order to design general-purpose sequential logic systems we must first partition the design into a *control algorithm*, which specifies the circuit operation, and an **architecture**, which **implements** the control algorithm. The control algorithm dictates how the circuit operates at an abstract level and may be independent of the hardware solution. When formulated correctly the control algorithm is translated into hardware to implement a design.

A formal method suited to the development of control algorithms for synchronous sequential circuitry is the *algorithmic state machine* (ASM) chart. An ASM chart is superficially a flowchart, but it expresses a concept of time intervals in a precise way, whereas a flowchart defines a sequence of events and not their time duration. An algorithmic state machine moves through a sequence of states, where each state is a unique position in a control algorithm, defined by the previous state and some status variables or inputs. For synchronous systems, state times are defined by the clock signal; active clock transitions cause a change from *present state* to *next state*.

The basic symbols for ASM chart notation are those already used for a state, denoted by a square box with a name attached to it, and a decision box (or conditional branch), where the route taken depends on the value of the status variable tested in the decision box.

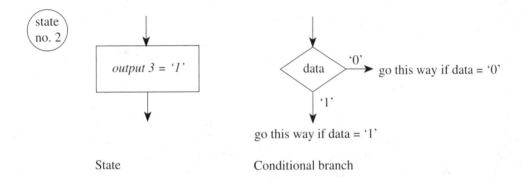

State   Conditional branch

Any value written inside the state box is an output generated unconditionally in that state; here any signal allocation written in *italics* denotes an output in that state. The control algorithm may be expressed as an ASM chart using these symbols. The circuit design is the translation of the ASM chart into hardware. As an example, the ASM chart for a 2-bit shift register is given by a sequence of states that are followed according to the input data. Each new state then depends on the value of a control input, and so each state is followed by a conditional branch. At each clock edge the system moves from one state, through the conditional branch, to a new state that depends on the input data. For a shift register, assuming all bistables start reset, the first state is with all bistables at '0'. If the following input data, at the next clock edge, is '0' the device remains reset. When a '1' is presented this will be latched by Q_A. If it is followed by a '1' the shift register fills with '1's, whereas if it is followed by a '0' the first '1' will be remembered in the second bistable, Q_B, and the '0' in the first, Q_A. The ASM chart is then

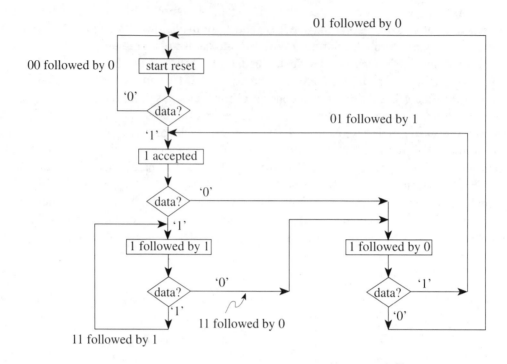

There are a number of ways to translate the ASM chart into hardware. Since we need to remember each state, it is possible to use a single bistable to remember each state. The position in the ASM chart sequence can then be determined from which bistable is set. This is actually called the *one-hot* solution. It is more efficient to *code* a state using a number of bistables. It is possible to **encode** 2^N states using N bistables, since each bistable is either set or reset in a particular state. This clearly reduces the number of bistables used but the penalty is that this approach can incur more combinational logic (this is inherent in using bistables to code each state). For example, an ASM chart with eight states can be implemented using 3 bistables; the outputs of the bistables need to be **decoded** to determine the present state. The outputs for each state are written above the box signifying that state; e.g. 01 written above a box signifies that for an order $Q_A Q_B$ then $Q_A = $ '0' and $Q_B = $ '1'. Q_A and Q_B are called the *state variables*, since their values describe particular states. For the 2-bit shift register the one-hot solution is excessively complex, whereas the coded version clearly follows the input data. The bistable state allocation is written beside each box as $Q_A Q_B$, where bistable A takes the input data and bistable B takes the Q_A output as its input data.

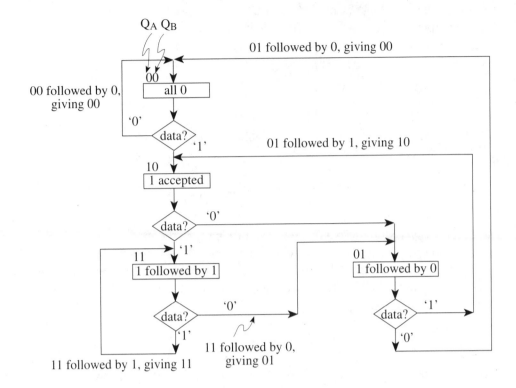

This ASM chart specifies the control algorithm that we require to implement. It specifies how each bistable is set or reset according to the input data and the previous state of the device. Design usually centres on implementation using edge-triggered D-type bistables. For each bistable, the state latched at an active clock transition depends on the data input and the current state, as defined by the bistable outputs. We then need to define the bistable data input D as a function of the input data and the current bistable outputs. This is achieved using a *present state/next state chart* which tabulates these inputs in a manner similar to a truth table. The transition between present and next state occurs at the clock edge. This is when the next state then becomes the present state of the device.

Often the full set of codes offered by a number of bistables is not needed to implement the full set of states. Should the ASM chart contain only five states, this will require three bistables at a minimum. The bistables offer eight combinations, so there will be three combinations, or states, which are not allocated: these are called *unused states*. It is usually better to assign unused states to enter the sequence, so that if they are entered inadvertently (via noise on the inputs, say) then the operation of the device will not be fully corrupted, since the sequence will return to that specified in the design. Unused states can be specified to rejoin the sequence at a point which confers a better minimised result. Better minimisation can also be achieved by changing the state allocations, that is, the codes used to represent each state. These points will be illustrated later; we will first consider the basic implementation.

The present state/next state chart for the 2-bit shift register is derived from the ASM chart as follows:

Present state Bistable outputs		Input data	Next state Bistable inputs	
Q_A	Q_B	I	D_A	D_B
0	0	0	0	0
0	0	1	1	0
0	1	0	0	0
0	1	1	1	0
1	0	0	0	1
1	0	1	1	1
1	1	0	0	1
1	1	1	1	1

These states cover all those possible for two bistables, so there are no unused states. The method of implementing this is to determine the conditions for which the bistable is set. This is directly equivalent to truth table extraction of combinational logic implementation. Bistable A is set for $D_A =$ '1'. There are four conditions for this, and these can then be minimised:

$$D_A = \overline{Q}_A \cdot \overline{Q}_B \cdot I + \overline{Q}_A \cdot Q_B \cdot I + Q_A \cdot \overline{Q}_B \cdot I + Q_A \cdot Q_B \cdot I = \overline{Q}_B \cdot I + Q_A \cdot I = I$$

Similarly, bistable B can be observed to be set after Q_A has been set, giving the expected extraction

$$D_B = Q_A$$

This gives the expected implementation, identical with that in Section 4.8.4. As with combinational logic implementation, it is often prudent to minimise logic extraction using a K-map. These tenets are best illustrated by application and we shall now design a basic sequential system, a counter.

4.9.2   Digital counter design – illustrating aspects of ASM design

4.9.2.1   Basic counter design – ASM charts in practice

A counter is a device that has a sequence of states that are related to count values. A counter can therefore be used to control the sequence of events in a system; at specified states, chosen events are activated or enabled, and the system is then controlled through the state sequence. Another application of a counter is to provide a delay of a specified number of clock cycles. This can give a very accurate method to provide a delay of a fixed time equal to the number of counted clock events.

Binary counters progress through a binary sequence and the length of this sequence is the *modulus*. This is similar to arithmetic systems where we also talk of modulus arithmetic. A modulo-8 counter has eight discrete states and is also called a *divide-by-8 counter*. In terms of a binary implementation, a divide-by-8 counter can be implemented using 3 bistables, since these can be used to encode the eight distinct states. The ASM chart for a divide-by-8 counter comprises eight states and the state machine progresses from state to state at each clock edge. In ASM chart design, the state assignment is arbitrary; any bistable combination can be chosen so long as it is unique for each state. In binary counters the coding is usually chosen to reflect a binary-coded count sequence. Three bistables are used to encode the eight states of a modulo-8 counter.

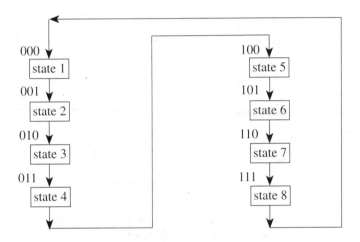

The present state/next state chart derived from this ASM chart is

Present state Bistable outputs			Next state Bistable inputs		
Q_A	Q_B	Q_C	D_A	D_B	D_C
0	0	0	0	0	1
0	0	1	0	1	0
0	1	0	0	1	1
0	1	1	1	0	0
1	0	0	1	0	1
1	0	1	1	1	0
1	1	0	1	1	1
1	1	1	0	0	0

The extraction for implementation is then, for D_A and D_B,

$$D_A = \overline{Q}_A \cdot Q_B \cdot Q_C + Q_A \cdot \overline{Q}_B \cdot \overline{Q}_C + Q_A \cdot \overline{Q}_B \cdot Q_C + Q_A \cdot Q_B \cdot \overline{Q}_C = \overline{Q}_A \cdot Q_B \cdot Q_C + Q_A \cdot \overline{Q}_B + Q_A \cdot \overline{Q}_C$$

$$D_B = \overline{Q}_A \cdot \overline{Q}_B \cdot Q_C + \overline{Q}_A \cdot Q_B \cdot \overline{Q}_C + Q_A \cdot \overline{Q}_B \cdot Q_C + Q_A \cdot Q_B \cdot \overline{Q}_C = Q_B \cdot \overline{Q}_C + \overline{Q}_B \cdot Q_C$$

D_C is set only when Q_C is LOW, and so

$$D_C = \overline{Q}_C$$

The circuit for a synchronous divide-by-8 counter is then

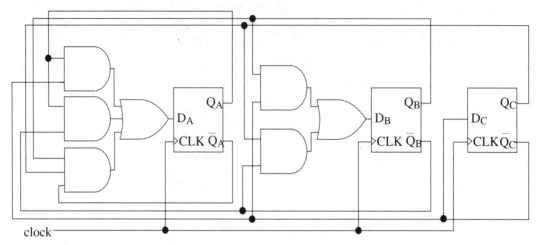

4.9.2.2 *Gray code counters – ASM implementation minimisation*

The state allocation can also reduce the combinational next state logic. A system commonly used is *Gray code*, where one bit only changes between adjacent states in the sequence. A Gray code for a 3-bit counter is:

$$000 \longrightarrow 001 \longrightarrow 011 \longrightarrow 010 \longrightarrow 110 \longrightarrow 111 \longrightarrow 101 \longrightarrow 100$$

These give eight distinct states for three bits, but do not reflect a binary-coded system. Should we need to determine a particular state, a decoder is required to interpret the bit pattern. For a counter, this Gray code can be used in the state allocation as follows:

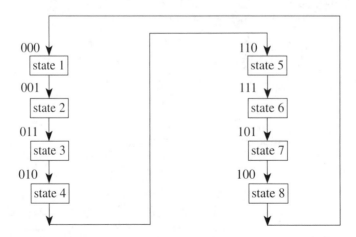

The present state/next state chart derived from this ASM chart is

Present state			Next state		
Q_A	Q_B	Q_C	D_A	D_B	D_C
0	0	0	0	0	1
0	0	1	0	1	1
0	1	1	0	1	0
0	1	0	1	1	0
1	1	0	1	1	1
1	1	1	1	0	1
1	0	1	1	0	0
1	0	0	0	0	0

Extraction for implementation for D_A, D_B, D_C is then, by minimisation,

$$D_A = Q_B \cdot \overline{Q}_C + Q_A \cdot Q_C$$

$$D_B = \overline{Q}_A \cdot Q_C + Q_B \cdot \overline{Q}_C$$

$$D_C = \overline{Q}_A \cdot Q_B + Q_A \cdot \overline{Q}_B$$

This is slightly simpler than the earlier extraction for a binary-coded counter since, even though the expression for D_C is more complicated, the implementation for D_A does not now require either of the two 3-input gates.

4.9.2.3 Further counter design – alternative sequences and unused states

We can also design counters which, for example, provide either divide by three or divide by five functionality, according to an input control signal. If this signal is '0' for a modulo-5 counter, and '1' for a modulo-3, we will call it $\overline{5}/3$. The ASM chart for this counter then contains five states, of which three are the states in the modulo-3 counter (when $\overline{5}/3$ is '1') and two are only entered in the divide-by-5 sequence (when $\overline{5}/3$ is '0').

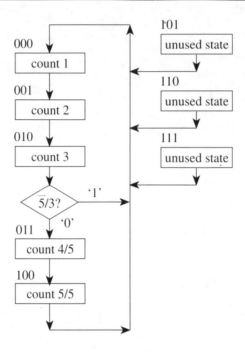

Three states are not used in the count sequences. In the ASM chart these can be allocated so that all bistables reset at the next clock exent, joining the sequence at the first count state. Should any of the unused states be entered accidentally, then the device will restart and correct operation will ensue. Unused states can actually be assigned for minimisation, as shown in Section 5.6.2.

Present state Bistable outputs			Count input	Next state Bistable inputs			
Q_A	Q_B	Q_C	$\overline{5}/3$	D_A	D_B	D_C	
0	0	0	X	0	0	1	First state of both sequences
0	0	1	X	0	1	0	Second state of both sequences
0	1	0	0	0	1	1	Continue with modulo-5 sequence
0	1	0	1	0	0	0	Go to the first state
0	1	1	X	1	0	0	Fourth state of modulo 5 count
1	0	0	X	0	0	0	Fifth and final state of modulo 5
1	0	1	X	0	0	0	Reset unused states
1	1	0	X	0	0	0	
1	1	1	X	0	0	0	

The unused states could enter the sequence at the point where the control signal is tested, or indeed at any point within the sequence. This will of course depend on the specification.

The extraction is more complicated in this example, so we will use a K-map to ease the minimisation process:

<table>
<tr><td colspan="5" align="center">D_A</td></tr>
<tr><td>$Q_A Q_B$ \ $Q_C 5/3$</td><td>00</td><td>01</td><td>11</td><td>10</td></tr>
<tr><td>00</td><td>0</td><td>0</td><td>0</td><td>0</td></tr>
<tr><td>01</td><td>0</td><td>0</td><td>1</td><td>1</td></tr>
<tr><td>11</td><td>0</td><td>0</td><td>0</td><td>0</td></tr>
<tr><td>10</td><td>0</td><td>0</td><td>0</td><td>0</td></tr>
</table>

<table>
<tr><td colspan="5" align="center">D_B</td></tr>
<tr><td>$Q_A Q_B$ \ $Q_C \overline{5}/3$</td><td>00</td><td>01</td><td>11</td><td>10</td></tr>
<tr><td>00</td><td>0</td><td>0</td><td>1</td><td>1</td></tr>
<tr><td>01</td><td>1</td><td>0</td><td>0</td><td>0</td></tr>
<tr><td>11</td><td>0</td><td>0</td><td>0</td><td>0</td></tr>
<tr><td>10</td><td>0</td><td>0</td><td>0</td><td>0</td></tr>
</table>

<table>
<tr><td colspan="5" align="center">D_C</td></tr>
<tr><td>$Q_A Q_B$ \ $Q_C 5/3$</td><td>00</td><td>01</td><td>11</td><td>10</td></tr>
<tr><td>00</td><td>1</td><td>1</td><td>0</td><td>0</td></tr>
<tr><td>01</td><td>1</td><td>0</td><td>0</td><td>0</td></tr>
<tr><td>11</td><td>0</td><td>0</td><td>0</td><td>0</td></tr>
<tr><td>10</td><td>0</td><td>0</td><td>0</td><td>0</td></tr>
</table>

The extraction thus reduces to:

$$D_A = \overline{Q}_A \cdot Q_B \cdot Q_C$$

$$D_B = \overline{Q}_A \cdot \overline{Q}_B \cdot Q_C + \overline{Q}_A \cdot Q_B \cdot \overline{Q}_C \cdot \overline{5/3}$$

$$D_C = \overline{Q}_A \cdot \overline{Q}_B \cdot \overline{Q}_C + \overline{Q}_A \cdot \overline{Q}_C \cdot \overline{5/3}$$

4.9.3 *Translating ASM charts into a software specification*

In order to translate an ASM chart into a software translation we must translate the present state/next state chart into software. Essentially we need to specify that if we are in a state then its position in the state sequence, and any tested inputs, define the next state in the sequence. The present state/next state chart comes from the ASM chart as does the software specification. The present state/next state chart includes all possible states and values of the inputs. These could be tested using multiple IF...THEN...ELSE statements but to reduce complexity we will use a CASE statement. The CASE construct requires variables to be tested, and these will be the current outputs of the bistables expressed in vector form. We then enter a new state, equivalent to assigning new values to each bistable according to any inputs tested in the current state. This assumes that the implementation is specifically targeted at D-type bistables. The assignment operator is what happens at a clock edge, namely that a new state is entered.

The present state/next state chart for a modulo 8 counter and its software specification are

Present state bistable outputs			Next state bistable inputs		
Q_2	Q_1	Q_0	D_2	D_1	D_0
0	0	0	0	0	1
0	0	1	0	1	0
0	1	0	0	1	1
0	1	1	1	0	0
1	0	0	1	0	1
1	0	1	1	1	0
1	1	0	1	1	1
1	1	1	0	0	0

```
;divide by 8 counter
CASE(Q[2..0])
   BEGIN

   #b000: Q[2..0]=#b001
   #b001: Q[2..0]=#b010
   #b010: Q[2..0]=#b011
   #b011: Q[2..0]=#b100
   #b100: Q[2..0]=#b101
   #b101: Q[2..0]=#b110
   #b110: Q[2..0]=#b111
   #b111: Q[2..0]=#b000
   END
```

and the software specification for the $\overline{5}/3$ counter derived from its present state/next state chart includes testing the $\overline{5}/3$ signal to determine which sequence the counter should enter and is

$\overline{5}/3$ Counter

Present state				Next state		
Q_2	Q_1	Q_0	$\overline{5}/3$	D_2	D_1	D_0
0	0	0	X	0	0	1
0	0	1	X	0	1	0
0	1	0	1	0	0	0
0	1	0	0	0	1	1
0	1	1	X	1	0	0
1	0	0	X	0	0	0
1	0	1	X	0	0	0
1	1	X	X	0	0	0

```
;Divide by 5 and divide by 3 counter
CASE (Q[2..0])
   BEGIN

   #b000: Q[2..0]=#b001
   #b001: Q[2..0]=#b010
   #b010: IF (5_3) THEN Q[2..0]=#b000
                   ELSE Q[2..0]=#b011
   #b011: Q[2..0]=#b100
   #b100: Q[2..0]=#b000
   #b101,
   #b110,#b111: Q[2..0]=#b000
   END
```

4.10 Further ASM chart design

Traditionally, an output generated in a particular state is written within the state symbol and the output will exist for the duration of that state. Outputs are often required that exist for a certain state, given that a certain condition is met. These are termed *conditional outputs*, since they depend on (are conditional on) the inputs, and are symbolised as follows:

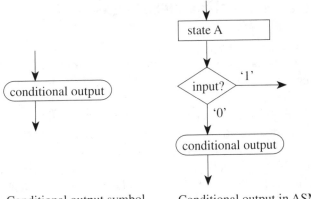

Conditional output symbol   Conditional output in ASM chart

Conditional outputs naturally occur in ASM charts after an input, or condition, has been tested. Given that we are in one state, A, then the conditional output will exist only if we are in state A and the input tested in state A is FALSE. The outputs that exist in a particular state are actually termed *unconditional outputs*.

There are alternative state machine descriptions, and these include the *Moore* and *Mealy* machines. These employ *state transition diagrams* which, like ASM charts, describe sequence of states in a sequential system. The traditional Moore machine is a state diagram containing only unconditional outputs and no conditional ones. On the other hand, the traditional Mealy machine has conditional outputs and a Mealy state machine can therefore have more than one output value in a state.

The states are symbolised using circles and the transitions are given by the links between the states. The state number and the output associated with that state are written within the circle denoting the state. The lines connecting states indicate the input conditions for a change between states, and the status of a particular input is written beside the line.

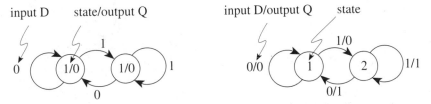

Moore state diagram for D-type flip-flop   Mealy state diagram for D-type flip-flop

In a Mealy state diagram (with conditional outputs) the lines connecting states have not only the inputs, but also the outputs. These diagrams can become rather complex for larger systems. Since the ASM method is a state machine employing both types of output, the ASM technique includes both cases and has therefore found increasing popularity for design purposes.

4.11 Asynchronous vs. synchronous design

We have so far considered *synchronous sequential design* only. In such a system all timing is specifically linked to the clock and the same clock event is used by all bistables. As a simpler explanation, everything changes at the same time and it can be inferred that this leads to greater reliability. There are also *asynchronous sequential systems* where timing is not specifically linked to the clock. Asynchronous systems move through a sequence of states, but outputs do not all change at the same time. The main advantages associated with asynchronous circuitry are a reduction in logic (and hence cost) which can confer an increase in speed. These advantages are only accrued with potential for unreliability and, since few engineers aim to build systems with inherent faults, it follows that asynchronous systems are normally used only in very special applications where their advantages are so important that a designer is willing to tolerate (and handle) any potential faults.

There are *asynchronous state machines* as well as synchronous ones. These are not synchronised to a clock but are driven by changes in the inputs, which cause changes between states. The design methods are more complex than for their synchronous counterparts and the results are more complex to analyze, but can achieve greater speed and minimisation in implementation.

The contrast between asynchronous and synchronous systems is perhaps best illustrated by counter design: asynchronous counters are simple but can be prone to error; synchronous counters are more complex and harder to design, but are more reliable. The asynchronous counters presented here are not asynchronous state machines, but are asynchronous systems that have bistable triggering signals that come from different sources.

An important element in asynchronous counter design is the *toggle-type* or *T-type* bistable. In a T-type, according to the status of a control signal, the outputs either toggle (as in the toggle state of a JK bistable), or latch.

T	CLK	Q_{n+1}	$\overline{Q}_{n+1}$
X	0	Q_n	$\overline{Q}_n$
X	1	Q_n	$\overline{Q}_n$
0	↑	Q_n	$\overline{Q}_n$
1	↑	$\overline{Q}_n$	Q_n

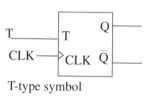

T-type symbol

This can be easily constructed using a D-type:

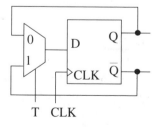

The T-type outputs change state on active clock transitions when the T input is HIGH, so an input clock frequency is then divided by two to give a divide by two or modulo-2 counter. For a negative edge-triggered T-type this is

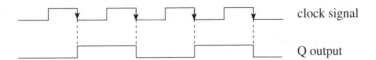

clock signal

Q output

If we connect more T-type bistables in series, connecting the clock for a following stage to one of the outputs of the previous stage, then at each stage we increase frequency division of the input clock frequency by two. Three bistables connected together in the following manner give frequency division by eight:

CLK

'1' — T Q_A '1' — T Q_B '1' — T Q_C

The timing diagram shows this frequency division in practice. If we **decode** the output states then we can see the counter cycling from 0 to 7 and back again:

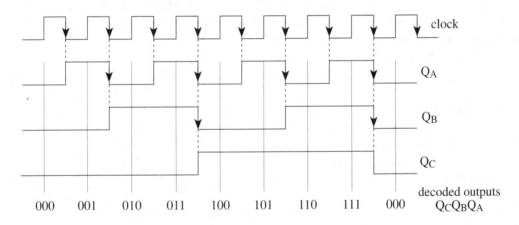

clock

Q_A

Q_B

Q_C

decoded outputs
$Q_C Q_B Q_A$

000   001   010   011   100   101   110   111   000

The system is, however, **asynchronous**, since not all clock signals are connected to the same clock input. Since the Q output of one stage provides the clock for the following stage, the second stage can only change two propagation delays after the clock, one propagation delay after the first output changes. This provides a clock that ripples through the system, and for this reason this counter is often called a *ripple counter*. Because of the ripple effect, the real timing diagram is then quite different from its idealised version:

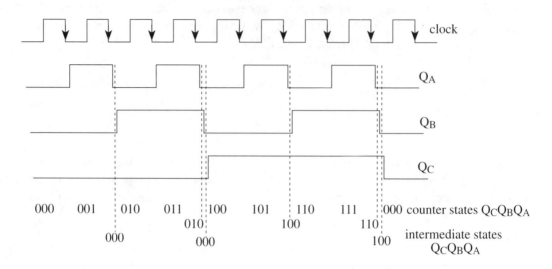

The decoded outputs are then $Q_CQ_BQ_A$, and state outputs and intermediate states are given above. The intermediate states are inherent in this design. Each bistable output provides the clock signal for the following stage, so the output of the bistable must fall before the next bistable changes its value. Since the outputs of counters are often decoded, for example, in sequence control, if a first device is enabled for a count of 2 (Q_C = '0', Q_B = '1', Q_A = '0') and a second device is to be enabled at the fifth cycle (Q_C = '1', Q_B = Q_A = '0'), then the first device will be enabled for a full clock period and then later for a very short time when the counter changes from Q_C = '0', Q_B = Q_A = '1' to Q_C = '1', Q_B = Q_A = '0'. This is usually completely unwanted and incorrect system operation will doubtless ensue. The second device (ostensibly only enabled for a full clock period at the fifth count cycle) will also be erroneously activated for a short time between the sixth and seventh clock cycles.

The main advantage of these asynchronous counters is **simplicity**. The asynchronous counter uses just T-type bistables, with no further combinational logic. This is much simpler than its synchronous equivalent. The price paid for this simplicity is in performance. Because of the unwanted intermediate states, a fully synchronous counter is usually preferred in sequential systems. There is one common application of a ripple counter. In digital watches the original clock frequency is provided by a quartz crystal, because the frequency can be controlled very precisely. (Quartz crystal clock circuits are described in Section 8.5.) Since the quartz crystal is physically very small, the clock frequency is very high (32.768 kHz) and needs division by a factor of 2^{15} to achieve a 1 Hz clock signal. This is often implemented using a ripple counter because of its simplicity (and hence low cost), and because the outputs do not need to be decoded to sequence control within the watch.

Inclusion of the asynchronous inputs to control the state of the bistable naturally leads to an asynchronous circuit. The ripple counter derived so far is only good to provide frequency division by 2^N for a system with N bistables. If we want a longer count sequence, then we can extend it by a factor of two by including a further bistable element. Earlier we designed a modulo-3 and a modulo-5 synchronous counter. These can be achieved in an asynchronous system by simply terminating the count, by resetting all bistables, when the counter reaches a specified *terminal count*.

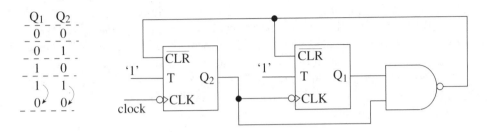

Count sequence   Asynchronous modulo-3 counter

If we require a modulo-3 counter then we want the system to reset when it reaches a terminal count of $Q_1 = Q_2 =$ '1'. We want to reset both bistables when this state is reached, so we need a clear signal derived from $Q_1 = Q_2 =$ '1'. Clear signals are usually active low, so if we NAND together Q_1 and Q_2 we obtain a signal that is LOW when both bistable outputs are HIGH. We can use this to force the count sequence to enter the state $Q_1 = Q_2 =$ '1' and then reset immediately to $Q_1 = Q_2 =$ '0', giving the required modulo-3 counter.

This circuit will usually work, but only just. This is because the first bistable to reset will cause the clear signal to return to HIGH, its inactive state. If there is a large difference in the *reset propagation times* (the time taken for the clear signal to take effect), then one bistable may not be reset. Usually, the clear signal is very short-lived, but sufficient to clear both bistables. Since operation is not guaranteed, it is therefore best to use a fully synchronous counter because it offers guaranteed performance with little extra combinational logic cost.

4.12   Concluding comments and further reading

This chapter has introduced sequential systems and devices. Though the edge-triggered D-type predominates in designs, some of the other bistables clarify various aspects of sequential system operation. Design also usually centres on synchronous sequential systems, since these offer greater potential for reliable operation than their asynchronous counterparts. ASM chart design is a major topic in synchronous sequential systems. This is a formal technique eminently suited to the description and implementation of synchronous sequential systems. As a design technique it is less suited to software implementation. Software implementation can provide a clear description of a synchronous sequential system, but with limited facility for the inclusion of asynchronous systems. Using software to specify systems now plays an important part in digital system design. The importance of software specification has increased concurrently with the range of devices that have been specifically developed for software specification: this is *programmable logic*, where software specifies a chip's function and the way that the logic circuits are connected together to achieve the software specification.

Most of the other texts on digital systems have a fairly clear coverage on bistables. See, for example, Wakerly, J. F., *Digital Design Principles and Practices* (Prentice-Hall, 2nd edn, 1994) and Unger, S. H., *The Essence of Logic Circuits* (Prentice-Hall, 1989). The classic text on ASM is Clare, C. R., *Designing Logic Systems Using State Machines* (McGraw-Hill, 1973), but it is hard to find a copy now. Prosser, D. and Winkel, F., *The Art of Digital Design* (Prentice-Hall, 2nd edn, 1989) is virtually dedicated to ASM design, but it is rather limited as an introductory text on bistables and logic. An expanded Moore and Mealy sequential machine classification can be found in Shaw, A. W., *Logic Circuit Design* (Sanders, 1993), which provides an alternative to the ASM approach.

4.13 Questions

1 Implement the switch debouncer circuit again for a single-pole, double-throw switch, but using RS bistables instead of $\overline{RS}$ bistables.

2 The following circuit aims to provide an output Q whose frequency is 15 MHz when the clock frequency is 30 Mhz:

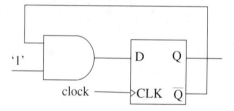

For the D-type bistable, $T_{SU} = 10$ ns, $T_H = 2$ ns, and $T_{PHL} = T_{PLH} = 13$ ns. For the AND gate $T_{PHL} = T_{PLH} = 12$ ns. Explain what happens in the clock cycle immediately following a rising clock edge. Will the circuit satisfy its aims?

3 Design a synchronous modulo-7 counter and implement it using positive edge-triggered D-type bistables and standard logic gates (AND, OR or NOT) only.

4 Design a synchronous modulo-7 counter for implementation using positive edge-triggered T-type bistables and standard logic gates (AND, OR or NOT) only.

5 Design an asynchronous modulo-7 counter and implement it using positive edge-triggered D-type bistables and standard logic gates (AND, NAND, OR, NOR or NOT) only.

6 Compare and contrast the implementation of a synchronous modulo-7 counter using positive edge-triggered D-types and T-types and an asynchronous modulo-7 counter using negative edge-triggered D-types.

7 The following two circuits are modulo-7 counters designed to provide two (HIGH) pulses of length 10 μs spaced by 20 μs and 30 μs, given a clock frequency of 100 kHz. Which of the circuits would you choose for a robust implementation?

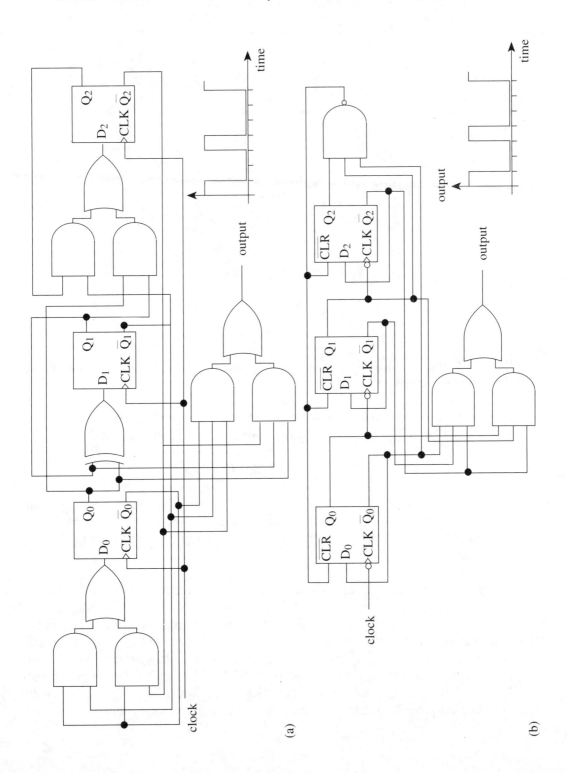

(a)

(b)

8 Express the following ASM chart in software using a CASE statement:

$Q_2Q_1Q_0$

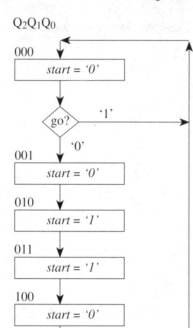

9 Express the ASM chart in the previous question, again in software, but using IF..THEN constructs only.

10 A system has four input channels A, B, C and D and one output channel. The output channel is selected from one of the input channels in the following sequence:

$$A,B,C,D,C,B,A,B,C,D,C, ...et \; seq.$$

Two control signals are provided: one to reset the system at any time to start with an output from channel A, the other is a short pulse synchronised to change the output selection. Design a circuit to implement this system and provide a circuit diagram for your design.

11 Design a system which operates either as a 2-bit shift register or as a 2-bit binary counter. The signal $S/\overline{C}$ is '1' to select action as a shift register, accepting a data input as Q_0 and shifting Q_0 to Q_1, and '0' to select the counter, where Q_1Q_0 count up in natural binary. Specify your design using an ASM chart and implement it using positive edge-triggered D-type bistables and standard logic gates (AND, OR or NOT). Given that for the bistables $T_{SU} = 10$ ns, $T_H = 2$ ns, and $T_{PHL} = 13$ ns and $T_{PLH} = 12$ ns, for any combinational gate $T_{PHL} = 11$ ns and $T_{PLH} = 12$ ns, estimate the maximum clock frequency at which your circuit will operate reliably.

12 Design a 3-bit bidirectional shift register that shifts its data left if a mode signal M is '0' and right if the mode signal is '1'. The shift register should accept data and shift on every third pulse of a clock signal. Specify the complete circuit using an ASM chart and implement it using positive edge-triggered D-type bistables and standard logic gates (AND, OR or NOT) only.

5 Programmable Logic Implementation

'Give us the tools and we'll finish the job.'

Winston Churchill

5.1 Programmable logic design

Implementation of digital circuits historically concerned finding a selection of chips that had the right functionality and the right attributes. Design was then a process of intimate knowledge and use of data books (and suppliers!). Nowadays programmable systems dominate design, though, often for reasons of cost, standard devices are still used where appropriate, including in particular buffers and registers. Programmable implementations require a language to specify the operation of the programmable logic device (PLD). There are many, often specific to a particular manufacturer, their programming (or **development**) system, or their device technology. The main programmable technologies are:

(a) *memory* – using (computer) memory chips to implement logic functions;
(b) *programmable array logic* – where fixed logic outputs are driven by programmable inputs;
(c) *programmable logic arrays* – where programmable inputs drive programmable outputs;
(d) *gate arrays* – which provide collections of uncommitted gates and the designer specifies how the gates are connected together; and
(e) *application specific integrated circuits* – where the designer uses gates with interconnection specific to a chosen application.

5.2 Programmable circuits

The main advantage of programmable architectures is that they can be configured to implement different functions, and this confers advantages at all stages of product development. Firstly, we can **stock** a smaller range of items (i.e. not the whole TTL family). Also, in **development** their use is advantageous since even when the wiring to the chip is fixed we can still change their functionality. Finally, in **production** we can still modify functionality in the light of manufacture and for some devices we can blow a *security fuse* that prevents anyone from finding out the contents. The advantages are considerable, thus contributing to their popularity.

In order for the circuits to be programmable we need some way of retaining data or connection. To remember data we can use a bistable mechanism, and some devices do this. For a level-triggered bistable, the clock signal can be thought of as a program signal to remember the data. We can also remember data by storing charge on a capacitor; a charged capacitor represents a '1' and a discharged one a '0'.

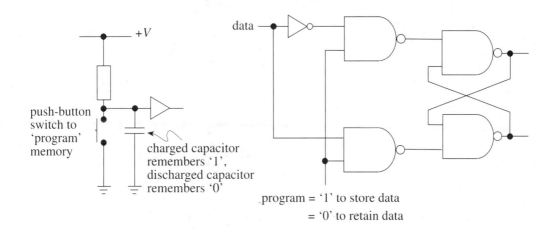

These circuits can be used to remember data that is stored within them. In programmable systems we often need to remember the connections to a logic function when the logic function or *architecture* of the device is fixed, but where the programming affects the way inputs and outputs are connected to the logic functions that the device offers. This can be implemented using *fuse logic*, where we blow a fuse to break a connection or leave it alone to keep the connection in place. The initial state of the device is with all fuses intact. The fuses control the connection between a logic input and the input to a gate. If we blow a fuse, or program it to be blown, then the logic input is disconnected from the input to a gate. If it is left alone then the connection will remain after programming. In this way we can program the inputs to the gates within a device. The fuses are often used in AND/OR arrays implementing sum-of-products expressions and we program the fuses to connect a variable (complemented or not) as an input to an AND gate.

In the following schematics, inverters are depicted as providing either a complemented (inverted) or an uncomplemented output. These two outputs from each inverter form an array of inputs which can be connected as an AND function whose output is fed to an OR gate. The AND function is made by blowing fuses; if the fuse is not blown then that input contributes to the AND function. We write a software specification which dictates which fuses should be blown to derive the required product expressions. In this manner we can choose to implement the EXOR function by connecting the functions $A\overline{B}$ and $\overline{A}B$ into the OR gate whose output is then EXOR.

In terms of programmable fuse logic
the fuses available for an AND/OR
configuration are

This is often represented in
manufacturers' databooks as

The result of programming them for
an exclusive OR function is

$\times$ = fuse, $\bullet$ = connection

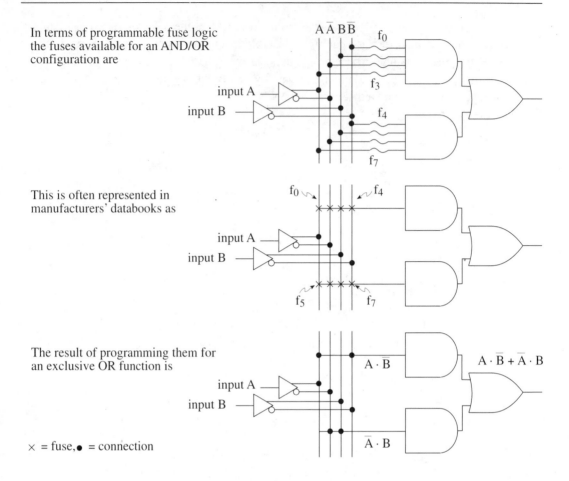

This gives the complete functionality for a logic function connecting two logic variables.
For this AND/OR configuration we blow fuses in the expression

$$\text{output} = (A+f_0)\cdot(\overline{A}+f_1)\cdot(B+f_2)\cdot(\overline{B}+f_3)+(A+f_4)\cdot(\overline{A}+f_5)\cdot(B+f_6)\cdot(\overline{B}+f_7)$$

where a blown link represents a '1'; so, for the fuses f = '0' represents an intact fuse and f =
'1' represents a blown one. To implement the exclusive OR function we blow fuses f_1, f_2, f_4
and f_7 to give

$$\text{output} = (A+0)\cdot(\overline{A}+1)\cdot(B+1)\cdot(\overline{B}+0)+(A+1)\cdot(\overline{A}+0)\cdot(B+0)\cdot(\overline{B}+1)$$
$$= A\cdot\overline{B} + \overline{A}\cdot B$$

There are many more architectures in programmable logic systems than AND/OR
structures. There are also devices that have a fixed architecture but where interconnection is
specified as part of the manufacturing process. This is again achieved by a software
description, but one that specifies the interconnections without using fuses. Semiconductor
devices are constructed from layers of material that form logic gates that are then connected
by a *metal layer* that is defined by a mask. The mask defines areas of metal that are retained

after fabrication and hence defines how the gates interconnect. Designers can specify these masks to manufacturers and for fast, economical, high-volume production this can be very attractive. In this manner the mask can achieve the programmable interconnection to the logic function.

5.2.1 Memory circuits

5.2.1.1 Memory functionality

The simplest entry point into programmable implementation is memory technology. Computer memory has slots in which we put the information that we want to remember. So that we can remember where we put it, we give each of the slots an address. The address is binary and n input *address lines* provide 2^n possible memory slots. Most memory chips are arranged to have more than one output, and are often configured to remember 4-bit *nibbles*, 8-bit *bytes* or even 16-bit *words*. The address lines select multiple outputs. The configuration of a memory chip is often implicit in its name. For example, 1024×4 implies a chip with 1024 slots each holding a 4-bit nibble. Note that 1024 is often condensed to 1K (i.e. a binary thousand) and we then have chips such as $2K \times 8$, which has 2048 slots for 8-bit bytes. By implication a chip with 1024 slots has 10 address lines (this gives an easy mnemonic since $2^{10} = 1024$), while one with 262144 slots requires 18 address lines. A widely-used $2K \times 8$ memory chip is the 2716, which is arranged as:

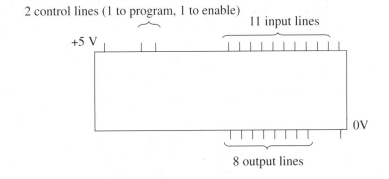

The 2716 contains 2^{11} (or 2048) 8-bit slots.

We can use memory to implement logic functions. It is clearly suited to combinational functions, though it can be used to provide next-state logic in sequential circuits (by combining inputs with the current state or value of the bistables' outputs to provide bistable inputs). The inputs are fixed and cover the entire address space of the device. This is identical to the truth table specification of a device, since the truth table tabulates the output for each input combination. These combinations then provide the address for a memory slot, and we insert the specified value of the output or function in each address.

The function of the memory can be interpreted in terms of an AND/OR description, which is common to manufacturers' descriptions. The AND plane determines the input combination that addresses a chosen location. This is fixed, since there is only one address for each location. The output is programmable, since it is the value of a logic function for a particular input combination. Memory is then often referred to as a **fixed AND/programmable OR** structure.

In this structure, the outputs can be connected low or high, and the OR structure is programmed to implement this. To implement a combinational function in memory we then need to transfer the output(s) specified in the truth table to become the content of the memory. The software that programs the logic function is then a list of output values for the contents of the memory, which are programmed into the memory using a development system. This can be a relatively inefficient implementation, since all states in the truth table are included, not just the minterms or maxterms alone (we might want minterms for an active high output or maxterms for active low).

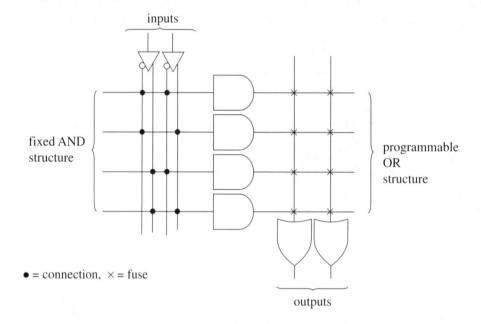

5.2.1.2   Types of memory

There are two main types of memory: *read-only memory* (ROM); and *random-access memory* (RAM). RAM is actually rather a misnomer, and a better description is *read/write memory*, since that is its function. RAM comes in two main varieties: *static* and *dynamic*. Dynamic memory essentially stores information using a capacitor that stores the data as charge. Charge leaks, and so the memory fades with time. If the capacitor is required to keep the data constant (to retain static data) then the memory needs to be *refreshed*. Static memory can use a bistable structure to hold data and therefore does not require any refresh circuitry. Using bistable structures means that static memory is more complex and therefore more expensive and potentially slower than its dynamic counterpart. Both types of RAM will lose their information when the power is switched off (though you can buy *battery-backed RAM* which incorporates a lithium cell to avoid this).

> *One major advantage of dynamic RAM is density,*
> *since much more dynamic memory can be accom-*
> *modated within a specified area. This is because*
> *the memory cell is much simpler than that for*
> *static RAM. Modern static RAMs are now as fast*
> *as dynamic ones.*

Since it remembers transient data only, a RAM is often described as *volatile*; battery-backed RAM is akin to ROM: it is *non-volatile* and will not lose its information. There are a number of varieties of ROMs. First, ROMs can be *mask-programmable,* where you specify the contents of the ROM to the manufacturer as a mask and they include it in production; alternatively, they are *field-programmable*, where you, as a designer, program the ROM on site. Some ROMs are *one-time-programmable* and can only be written to once. Many field-programmable chips can be reused, and there are *light erasable* types (usually ultraviolet light), though many modern ROMs are *electrically erasable* (EEROM). In EEROMs information is again stored as charge and a large voltage is used to obliterate the charge and clear the memory for reprogramming. *Flash* ROMs combine the advantages of EEROM and static RAM by offering a performance similar to static RAM when read, but with the facility for localised programming, rather than programming the whole device as in many ROMs. The choice between memory technologies is the inevitable compromise between cost and performance. In the development stages of product manufacture and as a student you will use mainly field-programmable EEROMs to reduce development cost. Choice in product manufacture is dictated by the same criteria, but with particular regard to the production volume.

5.2.2 Programmable array logic (PAL)

A memory-based solution is attractive when we want to implement an entire function. Usually, though, it is more efficient to implement either the minterms or the maxterms alone. We then implement an active high or active low output. This can be achieved using programmable array logic where the outputs are derived from an AND/OR function with functionality equivalent to a sum of products description of a logic circuit. The elements in a sum of products are called *product terms*. These determine the conditions for which the output is ON; otherwise it is OFF. Implementing a sum of products achieves a more efficient implementation than a memory implementation since we do not cover all input combinations.

The product terms are programmable through fuses, and we need to determine which fuses are to be blown via a software description. Since the product terms are an AND structure collected through an OR gate, the device is known as a **programmable AND/fixed OR** architecture. The inputs are available, inverted or not, to be connected via the fuses to the inputs to the OR gates. If the fuse is blown then the input is disconnected at that point.

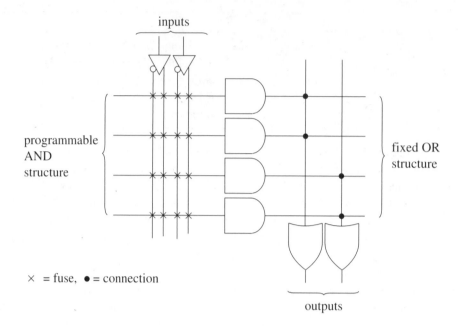

inputs

programmable AND structure

fixed OR structure

× = fuse, ● = connection

outputs

PALs can be combinational and implement combinational functions only. Some include bistables, which latch the outputs of combinational logic and this allows a PAL architecture to implement a state machine directly. Many PALs are fixed either as combinational or as sequential, whereas some are *output-configurable*, where a chosen output configuration (whether a particular output is combinational or derived from a bistable) is specified by the programmer. Combinational PALs essentially implement the sum of products logical format and in sequential PALs this forms the next state logic for an edge-triggered D-type. Should you require a different bistable functionality, this can be configured by using appropriate input combinations. The outputs can sometimes be enabled through an inverting or non-inverting buffer giving active low or active high outputs respectively. PALs usually offer a power-on reset and all bistable elements are cleared when the chip is first switched on.

> *Another major advantage of PALs, apart from compacting the circuit implementation, is timing. PAL architectures are fixed and offer fixed delay enabling timing to be analysed clearly.*

One basic combinational device is the 16L8, which is a combinational PAL available from many manufacturers. The first number in the chip's index (in this case 16) is the number of possible inputs that the chip can accept, while the second number (in this case 8) indicates the number of possible outputs that it can provide. Some of these inputs and outputs can be used either as an input or as an output when there is internal *feedback* from the output circuit to the AND/ OR array. The letter between these numbers indicates the type of PAL; in this case L indicates a combinatorial PAL with active LOW outputs. The active LOW output is derived from an inverter that can be switched ON or OFF, enabled or disabled respectively. This is actually tristate logic, which is described in Section 8.3. The basic architecture is a programmable AND/fixed OR structure as:

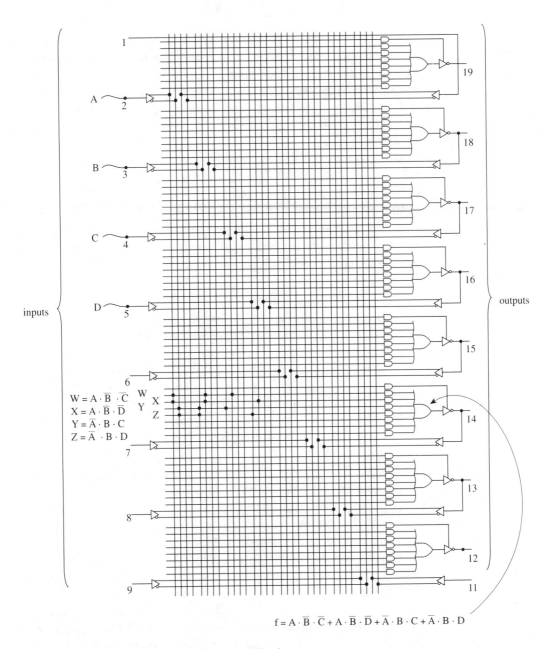

$$f = A \cdot \overline{B} \cdot \overline{C} + A \cdot \overline{B} \cdot \overline{D} + \overline{A} \cdot B \cdot C + \overline{A} \cdot B \cdot D$$

PAL 16L8 implementing John Wonderland's alarm

The fuse pattern implementing John Wonderland's alarm implements the function

$$f = A \cdot \overline{B} \cdot \overline{C} + A \cdot \overline{B} \cdot \overline{D} + \overline{A} \cdot B \cdot C + \overline{A} \cdot B \cdot D$$

You can see the connections in this circuit that form the product terms to produce the output signal. If our logic function was larger and contained too many product terms then it would quite simply not fit on the chip and we should have to find a chip large enough to accommodate our design. This might prove expensive, so we might sometimes interpret the specification differently, or alter the function slightly to meet design objectives at reduced cost with the smallest and cheapest device. In this case we could use a smaller device, since a lot of the PAL is unused.

For the present we shall consider systems where the outputs are permanently enabled. In some applications you might choose to enable the output only when a certain input condition is met. You can program an output to be enabled in that state by using the single-product term that provides the enable signal. If you want to use one of the pins (pin 12 to pin 18) which have feedback as an input, you switch the output inverter OFF and you can then use the pin as an input. In this manner there are sixteen possible inputs: ten are fixed inputs (and cannot be altered) and the remaining six can be provided by pins 12 to 18.

PLD device technologies have used bipolar, ECL and CMOS for the advantages that a particular logic technology confers. Early, bipolar PLDs were fast but suffered production difficulty. ECL PLDs are naturally very fast, whereas development of CMOS circuits took longer but now challenges the bipolar versions.

One basic sequential device is the 16R8. This has 16 possible inputs and 8 possible sequential outputs from bistable elements inside. It is then called a registered PAL, indicated by the R within its name. The registered elements are edge-triggered D-type flip-flops which have a common clock signal. The bistable outputs are fed back as inputs to the AND/OR arrays. There is a single control signal to enable the bistable outputs.

The edge-triggered D-types have a common clock which is connected to pin 1. This makes the 16R8 suitable for implementing synchronous sequential systems. The bistables' Q outputs are connected to the output pins via a tristate inverter controlled by pin 11, often called an *output enable* pin. The Q outputs are fed back to the AND/OR array and can be combined with the inputs to determine the next state of the system that the PAL implements. In this manner, the PAL can be programmed to implement a synchronous modulo-8 counter, described earlier.

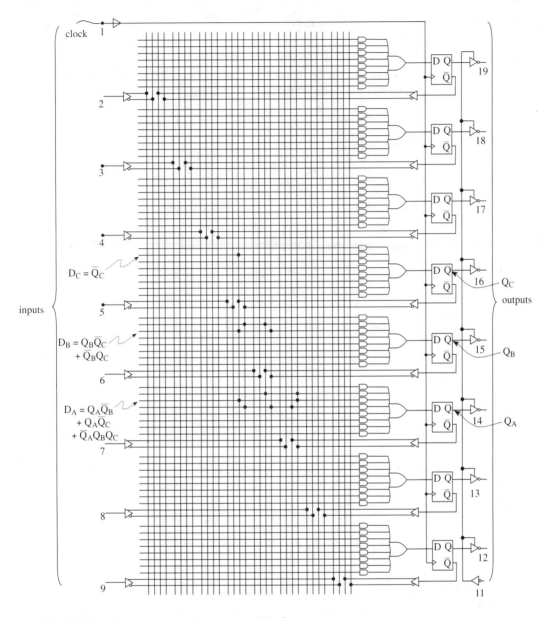

PAL 16R8 implementing synchronous modulo-8 counter

When it is implementing a synchronous modulo-8 counter, the device will have a fuse blown to implement the logic equations derived earlier:

$$D_C = \overline{Q}_C \qquad D_B = Q_B \cdot \overline{Q}_C + \overline{Q}_B \cdot Q_C \qquad Q_A = Q_A \cdot \overline{Q}_B + Q_A \cdot \overline{Q}_C + \overline{Q}_A \cdot Q_B \cdot Q_C$$

and the bistable outputs are fed back to be combined in the product terms to give the inputs to the bistables. We shall later consider techniques to derive these equations automatically and a software description to specify a *fuse map* that dictates exactly which fuses are blown.

5.2.3   *Programmable logic array (PLA)*

Though the PAL architecture is fairly versatile, it is limited by the fixed OR construction. A more versatile device, which can implement functions even more efficiently, is the programmable logic array (PLA), which has programmable AND (input) and OR (output) structures.

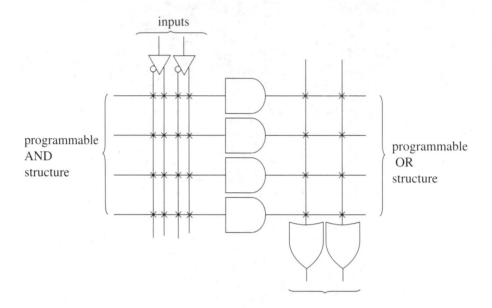

This can clearly compact the solution achieved by a PAL, since the outputs of the AND gates can be shared across OR gates. This allows much greater compaction to be achieved within a PLA. There is again a variety of combinational and sequential PLAs with a variety of output configurations. The implementation of this architecture is usually more costly, since it is larger and slower, and therefore for logic implementation a PAL is often more cost-effective.

5.2.4   *Gate arrays and application-specific integrated circuits*

Programming a PAL or PLA is not the end of the story. There are also *gate arrays*, which offer uncommitted collections of gates and latches, sometimes called 'seas of gates'. These are then committed to the circuit you have designed by the mask you specify, often using a computer-aided design (CAD) package. There are customised gate array circuits where the designer specifies the collection of gates prior to specifying their interconnection. The user can either use the manufacturer's sea of gates or latches and connect them up (or specify the mask to), or take pre-designed circuits and both place them and connect them together.

Field-programmable gate arrays (FPGAs) are an exciting development; the underlying architecture is fixed but their interconnection is programmable after power has been applied. The FPGA is then programmable in the same way as a PLD, but with a sea of gates with programmable interconnections. This avoids mask specification and the problems that it can cause. It also implies that by down-loading different software we can alter the functionality of the device. Since they can be very fast, this means that we could have an array of them inspecting car chassis for weld failures one day, and the next day updating the stock report! As ever, the advance of hardware has not been matched by a similar advance in software. There are still basic issues to be solved in appropriate data and program partitioning for implementation over an array of uncommitted gate arrays. Remember that, by the time you finish your course, software systems will be a lot better, but that hardware will have edged ahead as well!

The ultimate for many designers at present is the application-specific integrated circuit (ASIC). This is where you design the full circuit to implement a chosen function, the gates and their interconnections, all of which is then committed to silicon. There are many tools a designer will use here, not just logic design tools and simulators, but VLSI layout and placement tools also, since logic design at this level can impinge greatly on circuit implementation.

> *There are also sequencers and sequence-based PALs which have a slightly different architecture. Programmable technology moves very fast; the main elements presented in this text are commercially well proven and stable at the time of writing.*

5.3   The PALCE16V8 PAL and the PALASM language

Though there are many programmable device types, in this introductory text we deal with memory and PAL solutions only. This is an appropriate introductory level, since you will be able to see the benefits of programmable implementation without the additional complexity of the tools for design and synthesis of the more complex architectures. There are many manufacturers of PALs and many offer a software harness for you to program them with. In order to provide a wide introduction, we shall consider just one PAL and a single programming language. The PALASM 90 programming language is a very popular product, developed and marketed by Advanced Micro Devices Inc., and has been made freely available to universities and and colleges in the UK. It is a practical system for development and this text will centre on its use. The PALCE16V8 is an electrically reprogrammable PAL developed by AMD Inc. This is a versatile chip that can be arranged to implement many requirements. There are many other chips and many other languages; also, many manufacturers provide highly readable and lucid descriptions of their own devices and languages. An alternative PAL specification language, PLPL, and an alternative PAL, the PEEL 18CV8, are described in Appendix 2: these are used to re-implement the design examples in this chapter and show how a different language and a different PAL may be used. PALASM has many advantages over PLPL and there are more constructs catering for more advanced design methodologies. Remember that manufacturers promote their own devices; this is to your advantage since they highlight particular advantages of their systems which you can exploit in your designs.

5.3.1   The PALCE16V8 PAL

The PALCE16V8 is one of AMD's family of universal PAL devices with outputs that can be configured as either combinational or as sequential, derived from the output of a bistable. This is achieved using *macro cells*, which consist of user-programmable output logic elements. There are a number of chips in the family and they vary in size and in performance. The PALCE16V8 is CMOS and electrically erasable (the CE in its index); it has eight possible outputs derived from the macro cells, hence the V, for versatile, in its index. There are eight dedicated inputs, a clock signal and an output enable signal for registered PAL implementations. The arrangement connects chip inputs and feedback outputs to the macro cell using an AND/OR array

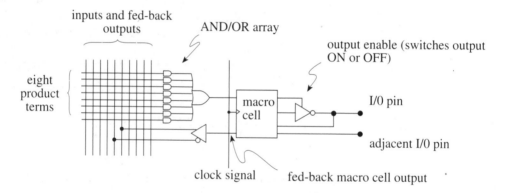

Each macro cell contains multiplexers that route specified signals through the device. These signals are derived from the inputs to the macro cells, which are the output of an AND/OR structure implementing a sum of products of the chip's inputs, and the fedback outputs. In the PALCE16V8 there are at most eight product terms feeding the OR gate. The OR gate's output is fed to become the PAL output or to become the input to a D-type bistable. This is how the macro cell implements combinational or registered outputs. The multiplexers' inputs are chosen according to the (software) specification; the multiplexers' control inputs reflect the required PAL implementation. To use a particular configuration, the user specification, which we shall write in PALASM, dictates the control signals which are fed to the multiplexers. The macro cell architecture is as follows:

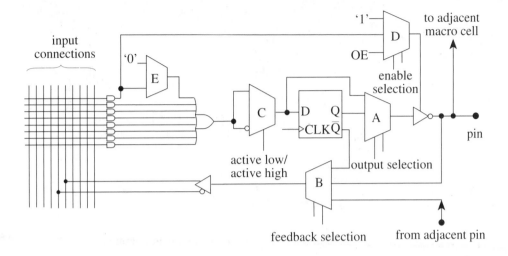

Multiplexer A controls the output selection and chooses whether the output is combinational (configured like a 16L8) or from a bistable (like a 16R8). Multiplexer B chooses the feedback selection, which can be derived from a combinational output or a bistable output, or the macro cell can be configured just to accept input from the adjacent pin. Multiplexer C controls whether the output is active high or active low. Multiplexer D controls whether the output is enabled directly, via an output enable input or via a product term. Multiplexer E removes the output enable product term from the sum of products when it is used as an output enable; otherwise it is included in the sum of product terms that appears at the output of the OR gate. The macro cells can be *bidirectional* and be used for both input and output (I/O) pins, or configured just to provide combinational input only. The signal from the adjacent pin is the input signal (via multiplexer B) when the macro cell is used purely as combinatorial input, and so each pin is fed to an adjacent pin. The specification of the PAL will state the required configuration that will be compiled to provide the necessary control signals for these multiplexers.

According to this architecture, the outputs can be selected, either active high or active low, registered (bistable) outputs derived from eight product terms with the output enabled separately. The outputs can also be (active high or active low) combinatorial input or output where the combinatorial output is derived from seven input product terms and the other product term can be used to disable the inverter to allow the pin to be used as an input. Purely combinatorial outputs are provided with the output inverter permanently enabled via V_{CC}, with output again active high or active low, and in this case derived from the full eight product terms. Finally, the macro cell can be configured to provide an input when the output inverter is disabled, via GND, thereby allowing the adjacent pin to be used as an input. This gives seven possible arrangements for a macro cell, three of which are for an active high output and three are for active low output. The first row in the following diagram gives the two registered arrangements. The second row shows committed combinatorial outputs, omitting the feedback and the bistables in the registered configurations. The third row includes feedback of a combinatorial signal but removes a product term to control the input/output selection. The final arrangement is for a pin which is used for input alone.

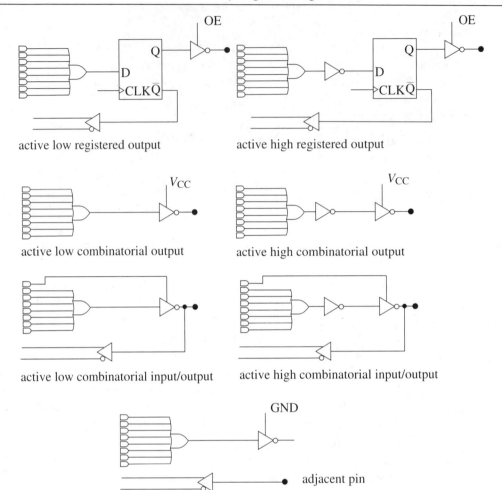

active low registered output   active high registered output

active low combinatorial output   active high combinatorial output

active low combinatorial input/output   active high combinatorial input/output

dedicated combinatorial input

The chip contains eight of these macro cells, which are connected to the committed inputs using the AND/OR array as follows:

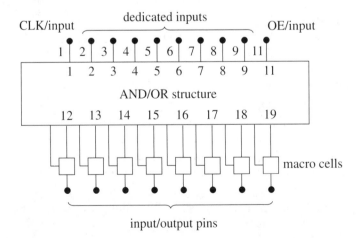

input/output pins

Pins 2 to 9 are combinatorial inputs to the AND/OR structures feeding the macro cells. Pins 1 and 11 are either combinatorial inputs or are the clock and output enable, respectively, if the macro cells are configured as bistables. The clock is positive edge-triggered; the output enable, OE, is LOW to switch the outputs on and HIGH to switch them off. The PALCE16V8 chip arrangement is then as follows:

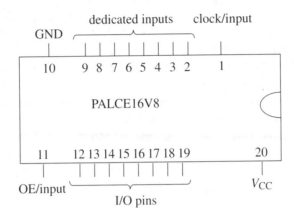

Since the chip is electrically reprogrammable the device can be reconfigured easily. There is also a security fuse that you can blow to keep the contents secret. One advantage of using the PALCE16V8 is that similar devices are made by a number of manufacturers.

5.3.2 PALASM

PALASM is a language to specify, and hence program, how PALs should operate and is used as the vehicle for software specification throughout this text. We first write a **specification** using the PALASM syntax. This is equivalent to the control algorithm in ASM design. This specification starts by first naming the chip and by specifying exactly which PAL it refers to, e.g. PALCE16V8. We then label each pin, saying how it will be used, and give it a software label and specify whether it is combinatorial or registered (a bistable output). We then define the functionality of the device in software. The language only relates logical variables and therefore implements a Boolean algebra only. There are no arithmetic functions in PALASM. The language allows logical operators including AND, OR, NOT and EXOR, which are symbolised by *, +, /, and :+: respectively. The statement `output = input_A*/input_B` specifies an output $= A \cdot \bar{B}$. These assignments can be conditional, so there are `IF...THEN...ELSE` constructs. These make use of `BEGIN...END` to ensure appropriate bracketing of commands. There are also `CASE` statements, which are particularly useful in sequential PLD descriptions.

The language is a specification language and **not** a programming language. This is actually quite a major difference from some programming languages, like C and BASIC, where the sequence of instructions dictates the order of events. All statements in a PALASM specification are 'executed' at the same time. This is the difference between a programming language that executes on a computer and a specification language for a PAL that has no central processing unit, just a 'simple' set of logic and latches.

There are keywords for *documenting* the design, keywords which are *statements* to express how the design works, *operators* to connect logical variables, and *simulation statements* to help us test how the specified design works. The PALASM language is ably described in the *PALASM 4 Reference Guide*, available from AMD (see Appendix 4), which gives a precise and thorough description. There is also *PALASM 4 Getting Started* which provides a tutorial introduction to the use of PALASM. A brief introduction only is given here, using commands that relate to many PAL devices. Some commands only relate to few devices and are omitted here. You should study the *PALASM 4 Reference Guide* for more advanced use and for many helpful comments.

All PALASM specifications described herein were developed with PALASM Version 4 generously provided by AMD Inc. Other PAL specification systems include *ABEL* (Advanced Boolean Expression Language) and *PLACE* (PEEL Logic Architectural Compiler and Editor).

The **statement keywords** in PALASM include

BEGIN-END	to bracket sets of instructions
CASE	to choose from an ordered list addressed by the CASE statement
CONDITIONS	to identify the start of state transition equations
DEFINE	to label statements for shorthand use
DEFAULT_BRANCH	to define a global default state transition
DEFAULT_OUTPUT	to define a global default output
ENABLE	to switch something on (usually the outputs)
EQUATIONS	to label the specification section containing Boolean equations
GND	logical '0'
GROUP	to collect several signals together
IF-THEN-ELSE	if some condition is met then execute the following instruction(s)
MEALY_MACHINE	for Mealy machine designs
MINIMIZE_OFF	to prevent Boolean equation reduction
MINIMIZE_ON	to enforce Boolean equation reduction
MOORE_MACHINE	for Moore sequential machines
PRESET	used to preload the device
RESET	used to clear the device
START_UP	to specify the state assumed in power up
STATE	to identify the state segment of Moore and Mealy sequential machines
STRING	to allow shorthand use of a string of characters
VCC	logical '1'
WHILE-DO	to execute a set of instructions while a condition is satisfied
.CLKF	to label a clock signal
.OUTF	to define Moore and Mealy state output equations
.RSTF	to define signal used for reset
.TRST	to enable outputs

The **operators** include

+	logical OR
*	logical AND
:+:	exclusive OR operator
:*:	exclusive NOR operator
( )	used to surround logic expressions
/	logical inversion, NOT
=	combinatorial assignment
:=	registered or state assignment
*=	latched assignment
->	local default
%	don't care
#b	binary
#d	decimal
#h	hexadecimal
#o	octal

The **punctuation marks** include

;	to precede comments
:	condition terminator in case statements
..	to indicates a range of values
,	to concatenate values and variables in statements
{ }	to substitute expressions
( )	surrounds logic expressions
[ ]	used in vector descriptions
Space	separator between variables and keywords

The **pin specifications** specify the chip you are using and how you use the pins on a chip.

CHIP	to name a PAL and specify its type
PIN	to specify how a pin is used
COMBINATORIAL	to label a combinational input or output
REGISTERED	to label a register output
LATCHED	latched output

The **simulation commands** include

CHECK	to verify that pin values equal expected values
CHECKQ	to verify that the registers' Q outputs equal expected values
CLOCKF	to generate a clock pulse in simulation
FOR-TO-DO	to repeat simulation sections
PRELOAD	to load specified values into registered outputs
SIMULATION	to start the simulation section

| TRACE_OFF | to specify end of traced segment |
| TRACE_ON | to specify which signals are recorded in a trace file during simulation |

The **documentation keywords** are

AUTHOR	to name the design's author
COMPANY	where the author works
DATE	when the design was started
PATTERN	used if the design conforms with a particular pattern
REVISION	to state how many times the device has been changed
TITLE	to state what the PAL does

The PAL specification is contained in a file with extension .PDS, e.g. WONDER.PDS, which is created using an editor. The specification is then compiled to produce a file that is the fuse map appropriate to our software description, compiling a file WONDER.PDS to produce a file WONDER.JED. This is the fuse map that we use to program our device. We can also simulate the device's operation (as worked out in the fuse map) to check that its operation is consistent with our expectations. The simulation segment is included with the device specification.

> *There are many design tools available to the modern engineer, including OrCAD for schematic logic circuit design. There is actually a mechanism within the PALASM system to include schematic capture via OrCAD.*

Before these concepts become too confused we shall now work through John Wonderland's alarm as a programmable design and implement it, first using memory, and then in a PAL.

5.4 Introductory programmable design example

We now implement John Wonderland's alarm using programmable logic. The minimised result used four 3-input AND gates, one 4-input OR gate and four inverters. In an SSI/MSI solution this implies four chips at minimum. The chip count can be reduced by using 2-input logic gates to replace the 3-input ones, as in the reduction by algebraic minimisation (see Section 2.10.1). We shall now use programmable logic to implement the alarm circuit within a single chip.

5.4.1   ROM implementation

The truth table specifies the contents of a ROM in terms of inputs and outputs. The inputs are the address of the data, and since a ROM programmer will work through in sequence, starting at address 0000 and working upwards, we only need to specify the output value as the content of the ROM at each location. This clearly depends on the programming environment. The six states for which the alarm is to be on, $f_1 \ldots f_6$, can be written in sum-of-products form as

$$f = \Sigma_{ABCD} \, (5, 6, 7, 8, 9, 10)$$

This could be then used to specify the ROM by writing to a file ROMFILE using software of the form

```
OPEN(ROMFILE)    ;Open a file for the ROM data
FOR i=0 TO 4     ;For the first five cases the alarm is off
  WRITE(ROMFILE,0) ;so write a 0 to the data file
FOR i=5 TO 10    ;It is on for the next six
  WRITE(ROMFILE,1) ;so write a 1 to the data file
FOR i=11 TO 15   ;The alarm is off for the last five cases
  WRITE(ROMFILE,0) ;so write a 0 (for each) to the data file
```

It would perhaps be easier to write this as a list of output values identical with the original truth table output definition, but it does not make for attractive copy! Note that there is no device ideally suited to implementing this function. Also, small, fast ROMs are a comparatively expensive solution.

5.4.2 Combinational logic PAL

5.4.2.1 PALASM description

We shall now implement the alarm circuit using a PALCE16V8. The first step is to translate the original specification into a PALASM description. The specification stated that we wanted the alarm to ring when Bo was with Chas or Dave, but not with Alice. The first stage of the PALASM description is to label the pins, four are to be used for a combinatorial input, one for a combinatorial output. The next stage is to determine the logic relations between these inputs and the output; this is then the PALASM program, contained in a file WONDER.PDS.

```
;PALASM Design Description
;********************
;Reference the design

TITLE    WONDER.PDS
PATTERN  A
REVISION 1.0
AUTHOR   M.S.Nixon
COMPANY  University of Southampton
DATE     05/01/94
```

```
;******************************
;now specify the pin connections

CHIP   wonder PALCE16V8 ;name and specify PAL type

PIN  1      alice   COMBINATORIAL    ; input
PIN  2      bo      COMBINATORIAL    ; input
PIN  3      chas    COMBINATORIAL    ; input
PIN  4      dave    COMBINATORIAL    ; input
PIN  15     alarm   COMBINATORIAL    ; alarm output
PIN  10     GND                      ; -ve supply
PIN  20     VCC                      ; +ve supply

;******************************
;specify the device's function

EQUATIONS

IF ((alice * /bo * (/chas + /dave)) +
   (/alice * bo * (chas + dave))) THEN
        BEGIN alarm = 1 END
     ELSE
        BEGIN alarm = 0 END
```

The first section of the specification is used for documentation. This is good practice since it gives you a way of recording how a device is built up. It will be omitted from later specifications, but only to save space in the text.

5.4.2.2   *Testing the program*

PALASM allows us to include functional testing within the program specification where the chip is simulated to check that it works. We then define a test sequence that specifies inputs and the expected output for that input combination. The simulator will then be used to test functionality. This is not an 'acid test'. That comes only when we connect the device up to real inputs to provide a real output. For a combinational device, the truth table can be used to specify the device's operation. The following code is then included with the original program:

```
;simulate it to check it works
SIMULATION
;follow the truth table entries
SETF /alice /bo /chas /dave ;set inputs to 1 or 0
check /alarm  ;check alarm is off for condition
```

```
SETF /alice /bo /chas dave   ;set next input value
check /alarm   ;check the alarm again
SETF /alice /bo chas /dave
check /alarm
SETF /alice /bo chas dave
check /alarm
SETF /alice bo /chas /dave
check /alarm
SETF /alice bo /chas dave
check alarm
SETF /alice bo chas /dave
check alarm
SETF /alice bo chas dave
check alarm
SETF alice /bo /chas /dave
check alarm
SETF alice /bo /chas dave
check alarm
SETF alice /bo chas /dave
check alarm
SETF alice /bo chas dave
check /alarm
SETF alice bo /chas /dave
check /alarm
SETF alice bo /chas dave
check /alarm
SETF alice bo chas /dave
check /alarm
SETF alice bo chas dave
check /alarm   ;all input combinations now tested
```

> *Design for testability concerns circuit function and differs from test vectors which are a user-specified test to check that performance is consistent with expectations of it. Design for testability is not considered in this text, but pointers for further study can be found later.*

5.4.2.3 *Result of compilation*

We can then compile the PALASM program that will work out a fuse plot to implement the chosen logic function. It first works out the minterms that implement the design and then minimises these to the prime implicants that provide a minimal implementation in an AND/OR configuration. After minimisation we have a file, WONDERLAND.JED, which can be disassembled to see the logic equations that the programmed PAL will implement. The disassembled file is WONDER.PL2 file, which contains the minimised equations for implementation together with the pin descriptions, which can be compared with the original result for the minimised alarm in Section 2.10.1.

```
TITLE        WONDER.PDS
PATTERN   .  A
REVISION     1.0
AUTHOR       M.S.Nixon
COMPANY      University of Southampton
DATE         05/01/94

CHIP   WONDER   PAL16V8

PIN 1 ALICE COMB
PIN 2 BO COMB
PIN 3 CHAS COMB
PIN 4 DAVE COMB
PIN 10 GND
PIN 15 ALARM COMB
PIN 20 VCC

EQUATIONS

    ALARM  =  /ALICE * BO * CHAS +  ALICE * /BO * /CHAS
           +  /ALICE * BO * DAVE +  ALICE * /BO * /DAVE
```

The alarm is driven by a logical combination of inputs that is the same equation as that by K-map minimisation (indeed, it is pretty similar to the PALASM specification). This is to be expected, since we have merely achieved the result in different ways. The whole process is self-documenting at each stage, and we have a full record of the design process. By inspection of the logic equations we can see that the alarm is a function of four product terms only and to implement it in a PALCE16V8 is a bit of a waste. Only four product terms drive one output, using little of the functionality available in the PAL.

The main point is of course to produce a file that is sent to the programmer to tell which fuses to blow. This is the JEDEC file, WONDER.JED, which shows those fuses that are blown (depicted with a 1), and those that remain (indicated by a 0). For a reconfigurable PAL this is naturally the current state of the PAL when programmed. The JEDEC file is then transferred to the programming system, which uses it to program the PAL. Part of the JEDEC file showing the fuse connections for Wonderland's alarm are

```
L1024 0110011111111111111111111111111*
L1056 1001101111111111111111111111111*
L1088 0110111101111111111111111111111*
L1120 1001111101111111111111111111111*
```

In this file each row corresponds to the input to an AND gate, while each column corresponds to an input (adjacent columns contain complemented inputs). This can be compared with the manufacturer's circuit diagram for the PALCE16V8, which shows the inputs to each OR gate and indexes the rows and columns. In this case the four rows correspond to the four product terms (the inputs to the OR gate) that constitute the alarm. The connections for the product terms were shown earlier for a PAL 16L8 in Section 5.2.2.

5.4.2.4   Result of simulation

We then simulate the device to test the functionality of the program. We use a simulator that derives the value of an output from a design for a particular input combination. Should the actual output disagree with the expected value specified in the simulation segment then the simulator program will mark it as an error in the result. This is contained in the WONDERLAND.HST file.

```
PALASM4 PLDSIM - MARKET RELEASE 1.5 (7-10-92)
 (C) - COPYRIGHT ADVANCED MICRO DEVICES INC., 1992


PALASM SIMULATION HISTORY LISTING


Title     : WONDER.PDS         Author   : M.S.Nixon
Pattern   : A                  Company  : University of Sou
Revision  : 1.0                Date     : 05/01/94


PAL16V8          Page : 1
                 ggggggggggggggggg
  ALICE    LLLLLLLLHHHHHHHH
  BO       LLLLHHHHLLLLHHHH
  CHAS     LLHHLLHHLLHHLLHH
  DAVE     LHLHLHLHLHLHLHLH
  GND      LLLLLLLLLLLLLLLL
  ALARM    LLLLLHHHHHHLLLLL
  VCC      HHHHHHHHHHHHHHHH
```

The alarm is on for the six expected input conditions. The design simulates correctly and so should work when it is made into a complete circuit by connecting the PAL inputs to the logic inputs.

5.5   Programmable sequential logic examples

We shall now implement the modulo-8 counter designed earlier using programmable logic. Though we can employ a ROM-based solution, this is rarely used. It is more common to use a PAL architecture.

5.5.1   Sequential ROM implementation

A ROM can be used to implement sequential functions. We cannot achieve a single-chip implementation, but must use chips for the bistables to accompany the ROM, since the ROM, even though it has memory, is only suited to combinational functions. It can be used within a level-triggered architecture, but edge-triggering is usually preferred in sequential systems. The potential role of a ROM is inherent in the general description of sequential systems.

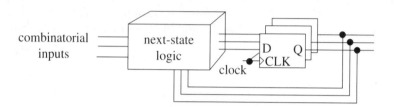

The ROM then implements a *look-up table* combining the current state and combinational inputs. This then determines the next state of the device. We then need to program the ROM with the present state-next state chart derived from the ASM design. For completeness, a modulo-8 counter designed earlier will require a ROM with 3 inputs and 3 outputs. For each address in the ROM we program a 3-bit word corresponding to the next state in the ASM chart.

5.5.2   Sequential PAL implementation

5.5.2.1   PALASM description

The difference between a sequential PLD and a combinational one is the introduction of a clock. In PALASM there is no difference, however, because all instructions are 'executed' at the same time since they specify a hardware connection. A simple introduction is to use a CASE statement where the conditionals for the CASE statement are the state variables and the current state and value of any input variables determine the next state of the system. At an active clock transition, the part of the CASE statement that relates to the current state will be used to determine the next state of the device. This is then identical to ASM design. After the active clock transition, the output is the new state registered by the bistable elements. The specification of the sequence of states in a modulo-8 counter is then a CASE statement with a list of all states and the next state.

```
;PALASM Design Description

TITLE     COUNTER.PDS
PATTERN   A
```

```
REVISION 1.0
AUTHOR    M.S.Nixon
COMPANY   University of Southampton
DATE      05/01/94


CHIP counter PALCE16V8 ;name and specify chip

;Define pin connections

PIN  1        clk       COMBINATORIAL   ; Clock input
PIN  12..14   z[3..1]   REGISTERED      ; (State) Outputs
PIN  11       oe        COMBINATORIAL   ; Output enable
PIN  10       GND                       ; -ve Power supply
PIN  20       VCC                       ; +ve Power supply


EQUATIONS ;specify how the counter operates

CASE (z[3..1])
  BEGIN ;define sequence of next states
    #b000: BEGIN z[3..1]=#b001 END
    #b001: BEGIN z[3..1]=#b010 END
    #b010: BEGIN z[3..1]=#b011 END
    #b011: BEGIN z[3..1]=#b100 END
    #b100: BEGIN z[3..1]=#b101 END
    #b101: BEGIN z[3..1]=#b110 END
    #b110: BEGIN z[3..1]=#b111 END
    #b111: BEGIN z[3..1]=#b000 END
  END
z[3..1].CLKF = clk


SIMULATION ;check sequence is as expected

SETF /clk  /oe          ;set clock low to start
                        ;and enable the outputs
CLOCKF clk              ;issue clock pulse
CHECK /z[3] /z[2] /z[1] ;check count 0 0 0
CLOCKF CLK              ;next clock pulse
CHECK /z[3] /z[2]  z[1] ;check count 0 0 1
CLOCKF CLK
CHECK /z[3]  z[2] /z[1] ;check count 0 1 0
CLOCKF CLK
CHECK /z[3]  z[2]  z[1] ;check count 0 1 1
CLOCKF CLK
CHECK  z[3] /z[2] /z[1] ;check count 1 0 0
```

```
CLOCKF CLK
CHECK   z[3]  /z[2]   z[1] ;check count 1 0 1
CLOCKF CLK
CHECK   z[3]   z[2]  /z[1] ;check count 1 1 0
CLOCKF CLK
CHECK   z[3]   z[2]   z[1] ;check count 1 1 1
CLOCKF CLK                 ;now go back to start
CHECK /z[3]  /z[2]  /z[1] ;check count 0 0 0
```

The CASE statement effectively executes in **parallel**. At each active clock transition the state (or present value of the bistable outputs) is found in the CASE statement look-up table and this then determines the next state of the outputs. The elements of the CASE statement are not executed in sequence; we have only written them in order for clarity. The clock is shown as common for all three registered elements using pin 1 as designated for a PALCE16V8.

The disassembled PAL equations are (omitting the pin specifications)

```
Z[1]    :=  /Z[1]
Z[1].CLKF  =  CLK

/Z[2]   :=  Z[1] * Z[2]
        +  /Z[1] * /Z[2]
Z[2].CLKF  =  CLK
/Z[3]   :=  /Z[1] * /Z[3]
        +  /Z[2] * /Z[3]
        +  Z[1] * Z[2] * Z[3]
Z[3].CLKF  =  CLK
```

These are identical to those achieved by ASM design (see Section 4.9.2.1).

5.5.2.3   Testing the program

There are two stages to testing the sequential PAL; the first is to check that the simulation sequence shows that the operation of the device is consistent with its design via simulation. Simulation uses the clock and c denotes a clock event in the simulation listing. The PALCE16V8 resets all register elements and hence the pin outputs are '1', HIGH (consistent with the inverter feeding the output pin), when power is applied. All specifications herein refer to pin values, rather than the internal PAL register states, as specified in PALASM. The simulation sequence starts with all pin outputs HIGH. The simulation actually starts by setting the clock LOW (this is mandatory in the PALASM system), and the output enable LOW, so that the state outputs can be observed on the pins. The sequence of simulation states is then, from the COUNTER.HST file:

```
        gc  c   c   c   c   c   c   c   c
CLK     LHHLHHLHHLHHLHHLHHLHHLHHLHHLHHL
GND     LLLLLLLLLLLLLLLLLLLLLLLLLLLLLLL
OE      LLLLLLLLLLLLLLLLLLLLLLLLLLLLLLL
Z[3]    HHLLLLLLLLLLLLLHHHHHHHHHHHHHHLL
Z[2]    HHLLLLLLHHHHHHLLLLLLLHHHHHHHHLL
Z[1]    HHLLLHHHLLLHHHLLHHHLLLHHHLLHHHLL
VCC     HHHHHHHHHHHHHHHHHHHHHHHHHHHHHHH
```

The device simulates successfully and propagates correctly through the state sequence. PALASM also provides a waveform diagram for diagrammatic representation of the output sequence. This allows you to check further the consistency of the designed operation of the device and shows how the signal at the outputs changes with time. It can be viewed, by selecting the appropriate option, on the computer monitor as

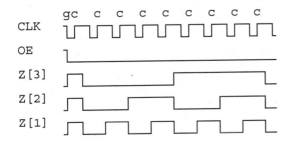

5.6 Design examples

We now study some design examples selected to highlight various aspects of design using PLDs. These concern the software coding, the relation to ASM-based design and aspects of implementation. These aspects concern the variety of possible software specifications, state allocation with particular reference to allocation of unused states, and minimising a design to fit within a PAL.

5.6.1 JK bistable

The JK bistable function has been defined earlier. Its main functions are to latch data for $J = K = '0'$, to set the bistable for $J = '1'$ and $K = '0'$, to reset it when $J = '0'$ and $K = '1'$, and to toggle (the outputs flip) for $J = K = '1'$. It is rarely used now, since its function can be obtained from a D-type bistable usually encountered in PAL architectures. One software description can use a CASE statement as

```
CHIP    JK_bistable  PALCE16V8
;JK bistable made from a positive edge-triggered D-type

;first define all active connections: clock, inputs and outputs
PIN  1   clk                   ;clock input
PIN  2   j  COMBINATORIAL      ;bistable input
PIN  3   k  COMBINATORIAL      ;bistable input
PIN  11  oe COMBINATORIAL      ;output enable
PIN  14  q  REGISTERED         ;bistable output
PIN  10  GND                   ;-ve power supply
PIN  20  VCC                   ;+ve power supply

EQUATIONS   ;specify operation
;specify function of J K bistable

CASE (j,k)
BEGIN
  #b11: BEGIN q=/q END ;toggle for j,k=1,1
  #b10: BEGIN q=1  END ;set for j,k=1,0
  #b01: BEGIN q=0  END ;reset for j,k=0,1
  #b00: BEGIN q=q  END ;latch for j,k=0,0
END
q.CLKF = clk



SIMULATION   ;check it works

SETF    /clk  /oe ;switch clock low to start simulation
                  ;and enable the outputs
SETF    j  k      ;set j = k = 1
CLOCKF  clk       ;issue clock pulse
CHECK   /q        ;test toggle
SETF    j  /k     ;set j = 1, k = 0
CLOCKF  clk       ;now check set state
CHECK   q
SETF    j  /k     ;keep j = 1, k = 0
CLOCKF  clk
CHECK   q         ;remain set
SETF    /j /k     ;go to latch
CLOCKF  clk
CHECK   q         ;latch set
SETF    /j k
CLOCKF  clk
CHECK   /q        ;test reset
```

This is then compiled yielding a D-type implementation which simulates successfully. The equations are (omitting the pin specifications)

```
Q   :=  Q * /K
    +  /Q * J
Q.CLKF  =   CLK
```

An alternative software description could employ the functionality of the JK bistable expressed using IF...THEN statements. This achieves the same result and compiles to the same solution, but achieves a description perhaps clearer to the designer.

```
CHIP    JK_bistable  PALCE16V8
;JK bistable made from a positive edge-triggered D-type

;first define all active connections: clock, inputs and outputs

PIN  1   clk                 ;clock input
PIN  2   j   COMBINATORIAL   ;bistable input
PIN  3   k   COMBINATORIAL   ;bistable input
PIN  11  oe COMBINATORIAL    ;output enable
PIN  14  q   REGISTERED      ;bistable output
PIN  10  GND                 ;-ve power supply
PIN  20  VCC                 ;+ve power supply

EQUATIONS   ;specify operation
;specify function of J K bistable
IF (j) THEN
        BEGIN
        IF (k) THEN BEGIN q=/q END ;toggle for j,k = 1,1
                ELSE BEGIN q=1   END ;set for j,k = 1,0
        END
        ELSE
        BEGIN
        IF (k) THEN BEGIN q=0   END ;reset for j,k = 0,1
                ELSE BEGIN q=q   END ;latch for j,k = 0,0
        END
q.CLKF = clk
```

When compiled, this achieves the same result as the earlier description. This result can be compared with the result achieved by using a present state/next state chart which for a JK bistable is

Present state	Next state
J K Q_n	Q_{n+1}
0 0 0	0
0 1 0	0
1 0 0	1
1 1 0	1
0 0 1	1
0 1 1	0
1 0 1	1
1 1 1	0

This can be minimised using a K-map as

	JK				
Q		00	01	11	10
0		0	0	1	1
1		1	0	0	1

This yields the same extraction as the PALASM implementations

$$D = Q \cdot \overline{K} + \overline{Q} \cdot J$$

The corollary is clear – **software specifications can be written many ways, each leading to the same implementation. The implementation should match that achieved by the ASM description if the specification is functionally identical**.

5.6.2   Johnson counter

The *Johnson counter* is a counter that can be used to enable processes in batches and is sometimes called a *twisted-ring counter*. The ASM description for a 3-bit Johnson counter is as follows:

$Q_1Q_2Q_3$

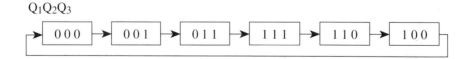

This has not specified values for the unused states. If the unused states are 'don't care' then the design result is that

$$D_3 = \overline{Q_1} \qquad D_2 = Q_3 \qquad D_1 = Q_2$$

This can be implemented using a shift register with the output fed back and inverted to become the input to the first bistable. Since the unused states are 'don't care', some unwanted states will propagate. If $Q_1Q_2Q_3 = 010$ then at the next clock cycle $Q_1Q_2Q_3 = 101$. If the counter enters either of these two states it will cycle between them indefinitely. PALASM offers a faster design method to study the effects of changing the unused states from 'don't care'. In the PALASM description we can include the unused states and, if the counter enters either state, it will then reset to all bistables cleared.

```
CHIP    johnson_counter   PALCE16V8
;pld for johnson counter


;first define all active connections: clock, inputs and outputs
PIN   1        clk
PIN   12..14   z[0...2]    REGISTERED     ;state outputs
PIN   11       oe          COMBINATORIAL  ;output enable
PIN   10       GND                        ;-ve power supply
PIN   20       VCC                        ;+ve power supply


EQUATIONS
;specify function using CASE statement implementing next state
CASE   (z[0..2])
BEGIN ;follow specified sequence
   #b000: BEGIN z[0..2] = #b100 END
   #b100: BEGIN z[0..2] = #b110 END
   #b110: BEGIN z[0..2] = #b111 END
   #b111: BEGIN z[0..2] = #b011 END
   #b011: BEGIN z[0..2] = #b001 END
   #b001: BEGIN z[0..2] = #b000 END
   #b101: BEGIN z[0..2] = #b000 END ;reset unused state
   #b010: BEGIN z[0..2] = #b000 END ;reset unused state
END   ;unused states could use OTHERWISE construct
z[0..2].CLKF = clk


SIMULATION ;test outputs as function of clock

SETF /clk /oe               ;set clock low to start
CLOCKF  clk                 ;issue clock pulse
CHECK   /z[0]  z[1]   z[2]  ;check 111 -> 011
CLOCKF  clk
CHECK   /z[0] /z[1]   z[2]  ;check 011 -> 001
```

```
CLOCKF   clk
CHECK    /z[0]   /z[1]   /z[2]   ;et seq.
CLOCKF   clk
CHECK     z[0]   /z[1]   /z[2]
CLOCKF   clk
CHECK     z[0]    z[1]   /z[2]
CLOCKF   clk
CHECK     z[0]    z[1]    z[2]   ;sequence complete
PRELOAD  z[0]   /z[1]    z[2]   ;101 -> 000
CLOCKF clk
CHECK    /z[0]   /z[1]   /z[2]   ;check that it resets
PRELOAD /z[0] z[1] /z[2]         ;010 -> 000
CLOCKF clk
CHECK    /z[0]   /z[1]   /z[2]   ;check the other
```

This then results in the compiled equations

```
/Z[2] := /Z[1]              /Z[1] := /Z[0]           Z[0] := /Z[2] * /Z[1]
       + /Z[2] * /Z[0]            + Z[2] * /Z[1]            + /Z[2] * Z[0]
Z[2].CLKF = CLK            Z[1].CLKF = CLK          Z[0].CLKF = CLK
```

This clearly does not approach the simplicity of the shift register solution. Even though it does actually fit within an 18CV8 PAL, it is possible to reduce the implementation to approach the simplicity of the shift register. This can be achieved by judicious allocation of the unused states. The first design reset all bistables when an unused state was entered. If we enter a different state from the reset state then it is possible for the design to become more compact. This can be achieved by studying the present state/next state chart or by trying new combinations at random. If the unused state $Q_1Q_2Q_3 = 101$ is set to rejoin the sequence with Q_1 as 1 then the Q_1 will be set in four states when Q_3 is LOW, thus simplifying the extraction for Q_1. This can be repeated for each unused state for each bistable, noting that the sequence should be rejoined at a valid state.

```
CHIP     johnson_counter   PALCE16V8
;minimised pld for johnson counter

;first define all active connections: clock, inputs and outputs
PIN   1         clk                    ;clock signal
PIN   12..14    z[0..2]   REGISTERED       ;state outputs
PIN   11        oe        COMBINATORIAL ;output enable
PIN   10        GND
PIN   20        VCC


EQUATIONS
;specify function using CASE statement implementing next state
CASE   (z[0..2])
```

```
BEGIN ;follow specified sequence
   #b000: BEGIN z[0..2] = #b100 END
   #b100: BEGIN z[0..2] = #b110 END
   #b110: BEGIN z[0..2] = #b111 END
   #b111: BEGIN z[0..2] = #b011 END
   #b011: BEGIN z[0..2] = #b001 END
   #b001: BEGIN z[0..2] = #b000 END
   #b101: BEGIN z[0..2] = #b110 END ;reset unused state
   #b010: BEGIN z[0..2] = #b001 END ;reset unused state
END
z[0..2].CLKF = clk


SIMULATION ;test outputs as function of clock

SETF /clk /oe ;set clock and enable low to start
CLOCKF   clk
CHECK    /z[0]   z[1]    z[2]
CLOCKF   clk
CHECK    /z[0]  /z[1]    z[2]
CLOCKF   clk
CHECK   /z[0]   /z[1]   /z[2]
CLOCKF   clk
CHECK    z[0]   /z[1]   /z[2]
CLOCKF   clk
CHECK    z[0]    z[1]   /z[2]
CLOCKF   clk
CHECK    z[0]    z[1]    z[2] ;sequence complete
PRELOAD /z[0]    z[1]   /z[2]
CLOCKF clk
CHECK     z[0]   z[1]   /z[2] ;check it resets
PRELOAD   z[0]  /z[1]    z[2]
CLOCKF clk
CHECK   /z[0]   /z[1]    z[2] ;check the other
```

The PAL specification now allocates the unused states differently. The simulation sequence tests these by using a PRELOAD command, which allows you to force the sequence to start at a chosen point rather than using the state sequence to reach that point. The PRELOAD command for the PALCE16V8 loads inverted values for the pin outputs, consistent with the chip's architecture. This specification results in

```
/Z[2]   :=   /Z[1] Z[2].CLKF   =   CLK
/Z[1]   :=   /Z[0] Z[1].CLKF   =   CLK
Z[0]    :=   /Z[2] * /Z[1]   +   /Z[2] * Z[0]   +   /Z[1] * Z[0]
         Z[0].CLKF   =   CLK
```

This is clearly a more compact implementation, since two of the bistables are identical to the shift register solution. The other bistable has an implementation that prevents progression of the unwanted states. Other allocations can similarly reduce the implementation. The corollary is clear – **unused states can be allocated for minimisation purposes**.

5.6.3   *Sequence detector*

A common requirement with serial digital data is to detect a particular sequence of bits within it. This is used, for example, in computer networks which communicate using serial data, where packets of information distributed around the network are preceded by a header that contains the address of the recipient. This header needs to be checked to determine whether the packet has reached its destination. Computer networks are actually much more sophisticated than this, but this does serve to illustrate a requirement to detect the occurrence of particular data streams within serial data.

We shall now design a circuit to detect the occurrence of the stream 000111 within a serial data stream. Assuming that the data is synchronised to a clock signal, we want to provide a signal that states that we have detected the sequence when it occurs. Given a data stream, we want to provide a logic signal that is HIGH when three successive '1's follow three successive '0's:

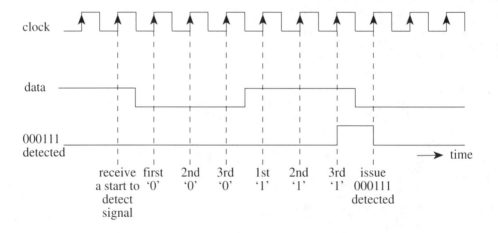

When we receive the first '0', we must remember it. If the following data bit is '0' as well, then we have had two '0's in succession and again we must remember it. If the second data bit is a '1' then we must restart, since the sequence has been violated. We then start again and wait for the first '0' in the sequence. Since we are aiming to detect a specified sequence 000111, i.e. three zeros followed by three ones, we restart the detection process whenever the sequence is violated. The signal showing that the specified sequence has been detected, *000111 detected*, should only be HIGH ('1') when three '0's have been detected in succession, followed immediately by three '1's. This can be summarised in an ASM chart as follows:

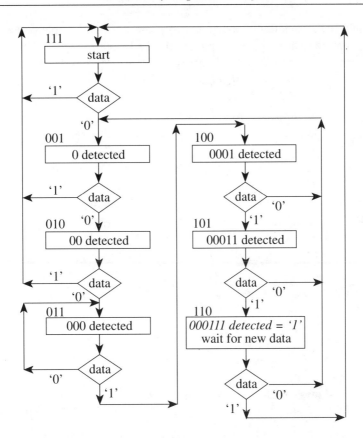

In order to implement this there are two possibilities. An intuitive design is to use a shift register-based solution. We can use the shift register to latch the input serial data and, when the specified pattern is stored within the register, we can provide the specified output signal. We then need to combine 6 outputs from a 6-bit shift register using an AND gate:

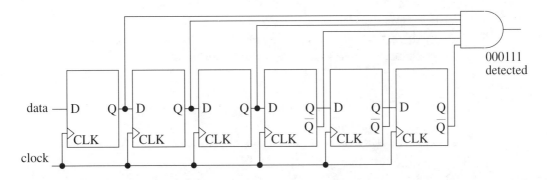

This can be implemented using PALASM, but it is rather wasteful in its use of bistables. Its main advantage is that the next-state logic is considerably less complex than that associated with a state-coded version using fewer bistables. There is a natural trade-off here since, even though the product terms given in a PAL can implement highly complex expressions, you might not want to use the remainder of the PAL. So why not just use the least complex solution? It is actually equivalent to a one-hot ASM implementation. A more efficient implementation can be achieved by coding the bistables. The state coding follows the ASM chart with binary-coded state allocations. Three bistables are needed, rather than the six associated with the one-hot implementation.

```
CHIP sequence_detector PALCE16V8
;A device to detect and count the occurrence of the sequence 000111
;within a serial data input stream
;first define the pins
PIN  1       clk                            ;clock input
PIN  2       data          COMBINATORIAL    ;input serial data
PIN  11      oe            COMBINATORIAL    ;output enable
PIN  12..14  states[2..0]  REGISTERED       ;states outputs
PIN  15      seq_detected  COMBINATORIAL    ;sequence detected
PIN  10      GND                            ;-ve power supply
PIN  20      VCC                            ;+ve power supply
EQUATIONS
CASE  (states[2..0])
BEGIN
  #b111: BEGIN
        IF (/data)
             THEN BEGIN states[2..0] = #b001 END;find first '0'
             ELSE BEGIN states[2..0] = #b111 END;'1' detected, restart
        END
  #b001: BEGIN
        IF (/data)
             THEN BEGIN states[2..0] = #b010 END;data = second '0'
             ELSE BEGIN states[2..0] = #b111 END;data = '1' so restart
        END
  #b010: BEGIN
        IF (/data)
             THEN BEGIN states[2..0] = #b011 END;data = third '0'
             ELSE BEGIN states[2..0] = #b111 END;data = '1' so restart
        END
  #b011: BEGIN
        IF (data)
             THEN BEGIN states[2..0] = #b100 END;find first '1'
             ELSE BEGIN states[2..0] = #b011 END;another '0',restart
        END
  #b100: BEGIN
        IF (data)
             THEN BEGIN states[2..0] = #b101 END;data = second '1'
             ELSE BEGIN states[2..0] = #b001 END;'0' so restart at 001
        END
  #b101: BEGIN
```

```
          IF (data)
                THEN BEGIN states[2..0] = #b110 END;data = third '1'
                ELSE BEGIN states[2..0] = #b001 END;'0' so restart at 001
          END
   #b110: BEGIN
          seq_detected = 1;output sequence detected
          IF  (data)
                 THEN BEGIN states[2..0] = #b111 END;'1', go to beginning
                 ELSE BEGIN states[2..0] = #b001 END;'0' so restart at 001
          END
    OTHERWISE: BEGIN states[2..0] = #b000 END ;reset unused state
END
states[2..0].CLKF = clk
SIMULATION
SETF   /clk   /oe       ;initialise clock and enable outputs
CHECK  states[2]  states[1]  states[0] /seq_detected
SETF   /data  ;follow expected sequence first
CLOCKF clk   ;find first '0'
CHECK  /states[2]  /states[1]  states[0]  /seq_detected
SETF  /data   ;set second '0'
CLOCKF  clk
CHECK  /states[2]  states[1] /states[0]  /seq_detected
SETF  /data   ;set third '0'
CLOCKF  clk
CHECK  /states[2]  states[1]  states[0]  /seq_detected
SETF  data    ;set first '1'
CLOCKF  clk
CHECK  states[2]  /states[1]  /states[0]  /seq_detected
SETF  data    ;set second '1'
CLOCKF  clk
CHECK  states[2]  /states[1]  states[0]  /seq_detected
SETF  data    ;set third '1'
CLOCKF  clk
CHECK  states[2]  states[1]  /states[0] seq_detected
SETF  data    ;found sequence so lets go to restart
CLOCKF  clk
CHECK  states[2]  states[1]  states[0]  /seq_detected
SETF  /data  ;input '0'
CLOCKF  clk
CHECK  /states[2]  /states[1]  states[0]  /seq_detected
SETF  data    ;now '1'
CLOCKF  clk  ;device should restart
CHECK  states[2]  states[1]  states[0]  /seq_detected
;now start half-way through sequence simulating 3 '0's detected
PRELOAD states[2] /states[1] /states[0]
```

```
SETF   /data
CLOCKF   clk ;should stay in state
CHECK  /states[2]   states[1]   states[0]   /seq_detected
SETF   data   ;now set a '1'
CLOCKF   clk ;should continue
CHECK  states[2]   /states[1]   /states[0]   /seq_detected
SETF   /data ;now set a '0'
CLOCKF   clk ;should go to '0' detected
CHECK  /states[2]   /states[1]   states[0]   /seq_detected
```

The OTHERWISE statement is now used in the CASE statement to specify the unused state. The waveform listing then shows that the correct signals have been derived:

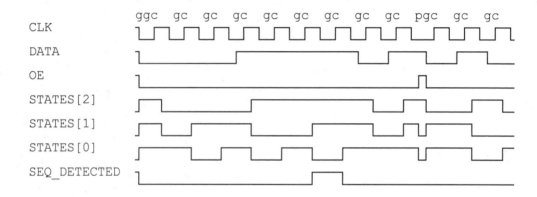

5.6.4   Data logger

The aim of this design example is to demonstrate the power of programmable design and to show how to make logic fit within an available PAL. The specification comes from an industrial design:

> A data logger is required that counts down when three successive '0's followed by three successive '1's within a stream of input data. There are four inputs: the start signal is high to start the data logging process. After the start signal has been high, three successive LOW values on an input data line should be logged while the start0 input is HIGH and the start1 input is LOW. After three successive '0's a start1 signal goes HIGH, then the start1 signal goes LOW, and three successive '1's should be logged (whilst start1 is HIGH and start0 is LOW). When the three successive '0's and the following three successive '1's have been logged then a counter should be decremented whilst start1 is HIGH. The device should only proceed to log three successive '1's after three successive '0's. These '0's should be logged in succession; if the start0 signal goes LOW whilst the three successive '0's are being logged, or if three successive '0's are not presented, the device should not proceed to the next logging stage but remain to

await three '0's. The device should only log three successive HIGH data values when the start1 signal is HIGH and start0 is LOW. The counter should decrement only after three successive '0's are followed by three successive '1's. The start0 and start1 signal are designed to overlap by the circuit presenting the data; should they not overlap, then the device should restart by awaiting the start signal.

The following timing signals illustrate the operation required of the data logger.

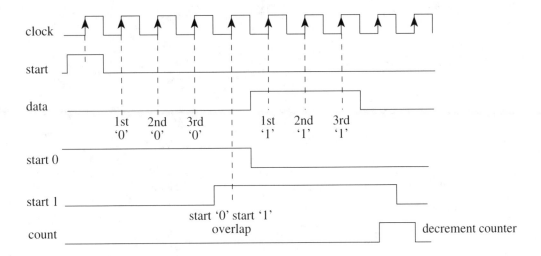

A written specification of this sort is very difficult to understand. A much clearer and more precise specification is the ASM chart.

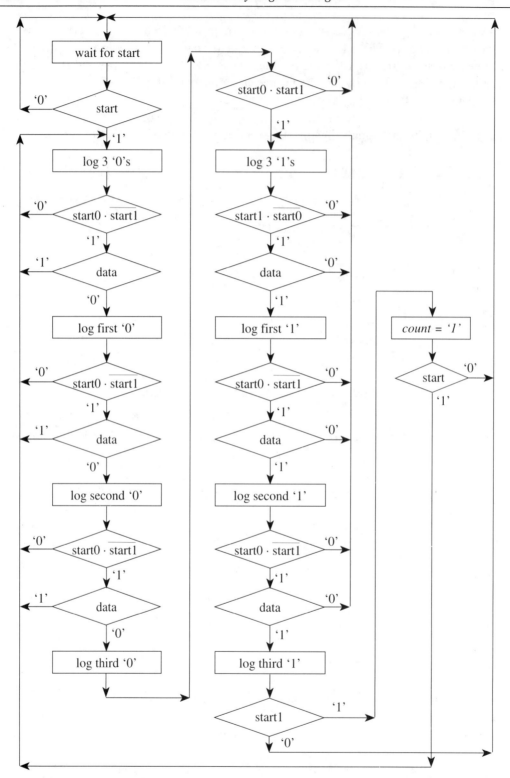

ASM chart for data logger

The counter here decrements and reduces by one each time the sequence is detected. It is known as a *down counter*. The whole of this design can be made to fit within a PALCE16V8, including the 3-bit synchronous counter. A direct software translation of the ASM chart is:

```
CHIP    data_logger  PALCE16V8
;A device to detect and count the occurrence of 000 then 111
;         in a serial data input stream
;First define the pins

PIN  1        clk                            ;clock input
PIN  2        data          COMBINATORIAL ;input serial data
PIN  3        start         COMBINATORIAL ;starts the data
PIN  4        start0        COMBINATORIAL ;starts the'0's
PIN  5        start1        COMBINATORIAL ;starts the '1's
PIN  11       oe            COMBINATORIAL ;output enable
PIN  12..15   states[3..0]  REGISTERED    ;state outputs
PIN  16..18   count[2..0]   REGISTERED    ;counter outputs
PIN  10       GND
PIN  20       VCC


EQUATIONS

CASE   (states[3..0])
BEGIN
  #b1111: BEGIN
          IF (start)
                THEN BEGIN states[3..0] = #b0001 END;check for 3 0s
                ELSE BEGIN states[3..0] = #b1111 END;wait for start
          count[2..0] = count[2..0]
          END
  #b0001: BEGIN
          IF (start0*/data*/start1)
                THEN BEGIN states[3..0] = #b0010 END;data = first 0
                ELSE BEGIN states[3..0] = #b0001 END;restart at 0s
          count[2..0] = count[2..0]
          END
  #b0010: BEGIN
          IF (start0*/data*/start1)
                THEN BEGIN states[3..0] = #b0011 END;data = second 0
                ELSE BEGIN states[3..0] = #b0001 END;restart at 0s
          count[2..0] = count[2..0]
          END
  #b0011: BEGIN
          IF (start0*/data*/start1)
```

```
                THEN BEGIN states[3..0] = #b0100 END;data = third 0
                ELSE BEGIN states[3..0] = #b0001 END;restart at 0s
            count[2..0] = count[2..0]
            END
#b0100: BEGIN
            IF (start0*start1)
                THEN BEGIN states[3..0] = #b0101 END;check for 3 1s
                ELSE BEGIN states[3..0] = #b1111 END;starts not 1, reset
            count[2..0] = count[2..0]
            END
#b0101: BEGIN
            IF (start1*data*/start0)
                THEN BEGIN states[3..0] = #b0110 END;data = first 1
                ELSE BEGIN states[3..0] = #b0101 END;restart at 1's
            count[2..0] = count[2..0]
            END
#b0110: BEGIN
            IF (start1*data*/start0)
                THEN BEGIN states[3..0] = #b0111 END;data = second 1
                ELSE BEGIN states[3..0] = #b0101 END;restart
            count[2..0] = count[2..0]
            END
#b0111: BEGIN
            IF (start1*data*/start0)
                THEN BEGIN states[3..0] = #b1000 END;data = third 1
                ELSE BEGIN states[3..0] = #b0101 END;restart
            count[2..0] = count[2..0]
            END
#b1000: BEGIN
            IF (start1)
                THEN BEGIN
                    CASE (count[2..0]) ;now decrement the counter
                      BEGIN
                      #b111: BEGIN count[2..0] = #b110 END
                      #b110: BEGIN count[2..0] = #b101 END
                      #b101: BEGIN count[2..0] = #b100 END
                      #b100: BEGIN count[2..0] = #b011 END
                      #b011: BEGIN count[2..0] = #b010 END
                      #b010: BEGIN count[2..0] = #b001 END
                      #b001: BEGIN count[2..0] = #b000 END
                      #b000: BEGIN count[2..0] = #b111 END
                      END
                    IF (start)
                        THEN BEGIN states[3..0] = #b0100 END;start at 0s
                        ELSE BEGIN states[3..0] = #b1111 END;restart
```

```
                        END
                ELSE BEGIN states[3..0] = #b1111 END;restart at beginning
        END
OTHERWISE:
        BEGIN
;for the unused states, check the start signal first and
;if the 0s are starting, go to that state
        IF (start)
                THEN BEGIN states[3..0] = #b0100 END
                ELSE BEGIN states[3..0] = #b1111 END
            count[2..0] = count[2..0]
        END
END
states[3..0].CLKF = clk
count[2..0].CLKF = clk
```

Note that the counter latches its value, except in the state where it is decremented. The main state coding follows the ASM chart directly. The simulation is not specified yet, since it is better to compile the design first to determine whether or not it fits on the chip. The result of compilation is failure, since one of the output pins requires too many product terms for the chip. states[1] actually requires nine product terms, whereas the device architecture allows only eight at most. This is actually documented as a reason for failure, and the JEDEC file is not produced. The equations can be disassembled, and for the offending bit the equation is

```
STATES[1]:= /START * /STATES[0] * /STATES[1] * /STATES[2]
        + /START * STATES[3]
        + /STATES[0] * /STATES[1] * STATES[2] * /STATES[3] *
/START1        + /STATES[0] * /STATES[1] * /STATES[2] *
STATES[3] * /START1        + /STATES[0] * /STATES[1] * STATES[2]
* /STATES[3] * /START0        + /STATES[0] * STATES[2] *
/STATES[3] * START1 * DATA * /START0        + /STATES[1] *
STATES[2] * /STATES[3] * START1 * DATA * /START0
        + /STATES[0] * STATES[1] * /STATES[2] * /STATES[3] *
/START1        * /DATA * START0
        + STATES[0] * /STATES[1] * /STATES[2] * /STATES[3] *
/START1        * /DATA * START0
    STATES[1].CLKF  =  CLK
```

This shows the nine product terms derived from the specification. It is possible to use Gray coding, as used before in counter design, to achieve a more minimised result. It is also possible to redefine the unused states, as illustrated in the Johnson counter design. This is not possible in this design, since there are no alternative allocations of the unused states. Varying the software description rarely achieves much in the way of minimisation; in this case, making the IF statements precede the CASE statement serves only to confuse the

software description. A Gray-coded implementation of this data logger appears the most
attractive option and this is achieved by using a Gray code to implement the states. The aim
of a Gray code is to cause as little possible change in bits between states, with the aim of
reducing the next-state logic; in this case it will serve to reduce the number of product terms
required by states[1] (essentially by spreading the load to other inputs) to make the
data logger fit within a PALCE16V8. An alternative approach is to use a chip with more
product terms, but this can be wasteful.

```
CHIP      data_logger  PALCE16V8
;A device to detect and count the occurrence of 000 then 111
;          in a serial data input stream
;First define the pins

PIN  1        clk                              ;clock input
PIN  2        data           COMBINATORIAL ;input serial data
PIN  3        start          COMBINATORIAL ;starts the data
PIN  4        start0         COMBINATORIAL ;starts the '0's
PIN  5        start1         COMBINATORIAL ;starts the '1's
PIN  11       oe             COMBINATORIAL ;output enable
PIN  12..15   states[3..0]   REGISTERED        ;state outputs
PIN  16..18   count[2..0]    REGISTERED        ;counter outputs
PIN  10       GND
PIN  20       VCC


EQUATIONS

CASE   (states[3..0])
BEGIN
   #b1111: BEGIN
           IF (start)
                THEN BEGIN states[3..0] = #b1110 END;check for 3 0s
                ELSE BEGIN states[3..0] = #b1111 END;wait for start
           count[2..0] = count[2..0]
           END
   #b1110: BEGIN
           IF (start0*/data*/start1)
                THEN BEGIN states[3..0] = #b0110 END;data = first 0
                ELSE BEGIN states[3..0] = #b1110 END;restart at 0s
           count[2..0] = count[2..0]
           END
   #b0110: BEGIN
           IF (start0*/data*/start1)
                THEN BEGIN states[3..0] = #b0111 END;data = second 0
                ELSE BEGIN states[3..0] = #b1110 END;restart at 0s
           count[2..0] = count[2..0]
           END
   #b0111: BEGIN
           IF (start0*/data*/start1)
                THEN BEGIN states[3..0] = #b0101 END;data = third 0
                ELSE BEGIN states[3..0] = #b1110 END;restart at 0s
           count[2..0] = count[2..0]
           END
```

```
     #b0101: BEGIN
              IF (start0*start1)
                    THEN BEGIN states[3..0] = #b1101 END;check for 3 1s
                    ELSE BEGIN states[3..0] = #b1111 END;starts not 1, reset
              count[2..0] = count[2..0]
              END
     #b1101: BEGIN
              IF (start1*data*/start0)
                    THEN BEGIN states[3..0] = #b1100 END;data = first 1
                    ELSE BEGIN states[3..0] = #b1101 END;restart at 1's
              count[2..0] = count[2..0]
              END
     #b1100: BEGIN
              IF (start1*data*/start0)
                    THEN BEGIN states[3..0] = #b0100 END;data = second 1
                    ELSE BEGIN states[3..0] = #b1101 END;restart
              count[2..0] = count[2..0]
              END
     #b0100: BEGIN
              IF (start1*data*/start0)
                    THEN BEGIN states[3..0] = #b0000 END;data = third 1
                    ELSE BEGIN states[3..0] = #b1101 END;restart
              count[2..0] = count[2..0]
              END
     #b0000: BEGIN
              IF (start1)
                    THEN BEGIN
                        CASE (count[2..0]) ;now decrement the counter
                          BEGIN
                          #b111: BEGIN count[2..0] = #b110 END
                          #b110: BEGIN count[2..0] = #b101 END
                          #b101: BEGIN count[2..0] = #b100 END
                          #b100: BEGIN count[2..0] = #b011 END
                          #b011: BEGIN count[2..0] = #b010 END
                          #b010: BEGIN count[2..0] = #b001 END
                          #b001: BEGIN count[2..0] = #b000 END
                          #b000: BEGIN count[2..0] = #b111 END
                          END
                        IF (start)
                        THEN BEGIN states[3..0] = #b1110 END;restart at 0s
                        ELSE BEGIN states[3..0] = #b1111 END;restart
                          END
                    ELSE BEGIN states[3..0] = #b1111 END;restart at beginning
              END
OTHERWISE:
              BEGIN
;for the unused states, check the start signal first and
;if the 0s are starting, go to that state
              IF (start)
                    THEN BEGIN states[3..0] = #b1110 END
                    ELSE BEGIN states[3..0] = #b1111 END
              count[2..0] = count[2..0]
              END
```

```
END
states[3..0].CLKF = clk
count[2..0].CLKF = clk


SIMULATION

SETF /clk /oe ;set clock low and enable outputs
SETF /start
CLOCKF
CHECK states[3] states[2] states[1] states[0]
CHECK count[2] count[1] count[0]
SETF start
CLOCKF
CHECK states[3] states[2] states[1] /states[0]
CHECK count[2] count[1] count[0]
SETF start0 /start1 /data
CLOCKF
CHECK /states[3] states[2] states[1] /states[0]
CHECK count[2] count[1] count[0]
CLOCKF
CHECK /states[3] states[2] states[1] states[0]
CHECK count[2] count[1] count[0]
CLOCKF
CHECK /states[3] states[2] /states[1] states[0]
CHECK count[2] count[1] count[0]
SETF start1
CLOCKF
CHECK states[3] states[2] /states[1] states[0]
CHECK count[2] count[1] count[0]
SETF /start0 data
CLOCKF
CHECK states[3] states[2] /states[1] /states[0]
CHECK count[2] count[1] count[0]
CLOCKF
CHECK /states[3] states[2] /states[1] /states[0]
CHECK count[2] count[1] count[0]
CLOCKF
CHECK /states[3] /states[2] /states[1] /states[0]
CHECK count[2] count[1] count[0]
CLOCKF
CHECK states[3] states[2] states[1] /states[0]
CHECK count[2] count[1] /count[0]
```

Here simulation is included, since the device compiles down to one which fits. The product terms are now split more evenly across the device by virtue of the Gray code implementation. This has actually been arranged so that bits change as little as possible from state to state. There are other implementations, and it is possible that these will lead to a more minimised result. It is usually the case that if the design can be made to fit within a specified chip with a suitable number of inputs and outputs; then, if the number of product terms can be made to fit within a chip architecture, minimisation goes no further, since there is no penalty associated with using extra product terms so long as the design fits within the chip.

Finally, we check the simulation to see that the device operates correctly. The waveforms are

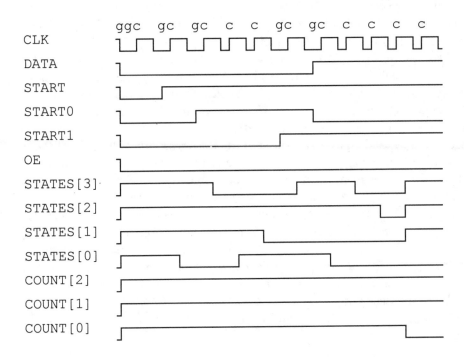

This is consistent with the required function of the device.

5.7 Concluding comments and further reading

In this chapter we have seen the basic programmable architectures now used to implement logic functions. The use of programmable architectures confers many advantages, not only because we can alter functionality without rewiring, but also because the specification is very clear and we now have complete documentation of the design process. However, we can only check that designs are consistent with their original objectives – there is no acid test except building the circuit.

The main vehicle for this text is PAL architectures. These offer a programmable implementation consistent with an introductory design level. Some design examples have shown that it is possible to design complex architectures within a single chip. The PALCE16V8 will be the main PAL architecture used in this text because of its versatility. There are many other PAL architectures, and some design packages can optimise implementation for parameters such as speed and cost. The main language used here is PALASM, and again there are many more languages. PALASM is sufficient at an introductory level and serves to highlight the advantages that programmable implementations can accrue. An alternative PAL and language combination is to use a PEEL18CV8 and the PLPL language and this is described in Appendix 2, using the same examples. Alternative implementations using PALASM, in particular the use of a Moore machine, appear later in the text.

Bostock, G., *Programmable Logic Handbook* (Butterworth-Heinemann, 2nd edn, 1993) describes PLD technology, with design examples, but without including introductory design methods. This includes a cross-section of technology associated with PLDs, as does Bolton, M., *Digital Systems Design with Programmable Logic* (Addison-Wesley, 1990) together with many further references on PLD developments and a review of device technology available at the time of writing. Many digital electronic texts provide a brief introduction to programmable circuits: Haznedar, H., *Digital Microelectronics* (Benjamin, 1991) reviews the insides of programmable circuitry, as do Weste, N. and Estraghian, K., *CMOS VLSI Design* (Addison-Wesley, 2nd edn, 1992). Haskard M. R., *An Introduction to Application Specific Integrated Circuits* (Prentice-Hall, 1990) gives an introduction to gate arrays and ASIC technology. There are many manufacturers of memory circuits that provide data books (see for example, National Semiconductor, Motorola, and Texas). Many manufacturers of programmable devices provide clear information on design, together with application examples; have a look at Advanced Micro Devices's *PAL Device Data Book* in particular, but also Intel's *Programmable Logic*, Philips's *Integrated Fuse Logic*, National Semiconductor's *Programmable Logic Devices Data Book and Design Guide*, Altera's *Data Book* (Altera is a specialist PLD manufacturer) or Cypress's *CMOS Data book* (for memories and PLDs). For gate arrays, the Xilinx *User Guide and Tutorials* manual provides a good introduction. As with integrated circuits, many vendors can supply datasheets on the products they sell.

5.8 Questions

1 Specify a 3-8 line decoder using PALASM and check your solution.

2 Specify an 8-1 line multiplexer for implementation in a PAL and check your result in part by comparing it with the solution to Question 1.

3 Repeat Question 2 on page 46 via a PAL specification and check your result.

4 Repeat Question 4 on page 46 via a PAL specification and check your solution.

5 Complete Questions 8 and 9 on pages 46 and 47 as a full PAL specification and check your solution.

6 Specify a divide-by-3 counter using the PALASM format.

7 Specify a 2-bit bidirectional shift register for a PAL and check that your solution is correct.

8 Complete Questions 8 and 9 on page 126 as a full PALASM specification and check that they work.

9 Implement the arithmetic functions specified in Question 12 on page 47 using a PALASM specification and check that you achieve the same solution as before.

10 Express a modulo-6 counter as a PAL using a state sequence encoded in natural binary. Investigate how Gray-coded state allocation and better unused state allocation can be used to minimise the implementation.

11 Express the problem of the shift register/counter in Question 11 on page 126 as a PAL specification and check that compilation delivers the same result as that question.

As mentioned in the Preface, the solutions can be obtained by anonymous FTP to ECS.SOTON.AC.UK where you LOGIN as anonymous, using the same password (reply anonymous to the Name and Password prompts), change directory to pub/digits (by cd pub/digits) where you will find a zipped file, digits.zip, containing source code for the design examples and the solutions to the programmable logic questions. Change the transmission type to binary (type bin) and obtain the file (type get digits.zip). This file needs to be unzipped, so use PKUNZIP (use as PKUNZIP digits.zip [destination directory]).

6 Number Systems, Coding and Arithmetic

'The scientists had another idea which was totally at odds with the benefits to be derived from the standardisation of weights and measures; they adopted to them the decimal system. ... Nothing is more contrary to the organisation of the memory and of the imagination. ... It's just tormenting the people with trivia.'

Napoleon Bonaparte

6.1 Numbers and coding systems

6.1.1 Common number coding systems

The basic form of a number is a collection of digits arranged into *integer* and *fractional* parts. The digits are arranged in order of significance, and numbers are *left justified* when the most significant digit is to the left. The arrangement of a number is

$$(D_n \times r^n + D_{n-1} \times r^{n-1} + \cdots + D_1 \times r + D_0).(D_{-1} \times r^{-1} + D_{-2} \times r^{-2} + \cdots D_{-m} \times r^{-m})$$

The part of the number to the left of the (decimal) point is an $n + 1$-digit integer and the part to the right is an m-digit fraction. The base of the number is called its *radix* and this is denoted by r in the number. Each digit is multiplied by the radix raised to the power associated with that digit. The least significant digit in the integer part is actually $D_0 \times r^0$, and the radix part is omitted since $r^0 = 1$. In order to avoid confusion, the radix is often used as a subscript to denote the number system used, so 11_{10} denotes eleven in the decimal system, whereas 11_2 denotes a binary number.

A *decimal number* is base 10 and so the radix r is 10. The decimal system uses ten symbols, 0, 1, 2 ... 9. The number 1874.28_{10} is then

$$(D_3 \times r^3 + D_2 \times r^2 + D_1 \times r + D_0).(D_{-1} \times r^{-1} + D_{-2} \times r^{-2} = (1 \times 10^3 + 8 \times 10^2 + 7 \times 10 + 4) + (2 \times 10^{-1} + 8 \times 10^{-2})$$
$$= (1 \times 1000 + 8 \times 100 + 7 \times 10 + 4) + (2 \times 0.1 + 8 \times 0.01)$$
$$= 1874.28_{10}$$

The *binary system* has only two values. The radix is two and the two symbols are 0 and 1. Each digit in a binary number is known as a *binary digit* or *bit*. In *natural binary*, the number 1101.01_2 represents the number

$$(D_3 \times 2^3 + D_2 \times 2^2 + D_1 \times 2^1 + D_0).(D_{-1} \times 2^{-1} + D_{-2} \times 2^{-2}) = (1 \times 2^3 + 1 \times 2^2 + 0 \times 2^1 + 1) + (0 \times 2^{-1} + 1 \times 2^{-2})$$

which in decimal is

$$= (1 \times 8 + 1 \times 4 + 0 \times 2 + 1).(0 + 0.5 + 1 \times 0.25)_{10}$$
$$= 13.75_{10}$$

Binary coded decimal (BCD) uses decimal digits coded using the binary system. The codes for each digit follow the binary sequence, starting with $0_{10} = 0000_2$, $1_{10} = 0001_2$, $2_{10} = 0010_2$ through to nine, which is represented by

$$1001_2 = (1 \times 8 + 0 \times 4 + 0 \times 2 + 1 \times 1)_{10}$$

Note that we need four bits for each BCD digit, since this allows us to code the ten digits. Six of the codes possible for the four bits (1010_2 through to 1111_2) are unused in BCD. This system has been popular in calculators.

This system is sometimes called 8421 BCD, owing to the natural binary weights used. 2421 BCD weights the three least significant digits using natural binary, but the MSB is only weighted as two, not eight as is 8421 BCD.

Octal is a decimal-coded version of a radix-8 binary system. According to the radix, eight digits are used, and these are 0, 1, 2, 3, 4, 5, 6, 7, the first eight decimal symbols. Each digit represents a collection of three binary digits, which are coded using decimal symbols. There is a type definition of octal in the C^{++} programming language, but the system is less used now, though it found early popularity in minicomputer systems.

Hexadecimal is a widely used system of radix 16 employing encoded binary numbers. Since binary numbers represent a large collection of bits, it is often convenient to compact them using a code. The hexadecimal system encodes four bit nibbles and, since a pattern of four bits has sixteen possible values, we need symbols for sixteen digits. Decimal symbols are used to represent the first nine of these, so

$$0_{16} = 0000_2, 1_{16} = 0010_2, 2_{16} = 0010_2 \dots 9_{16} = 1001_2$$

The first six letters of the alphabet are used to represent the remaining six combinations: $A_{16} = 1010_2$ (10_{10}); $B_{16} = 1011_2$ (11_{10}); $C_{16} = 1100_2$ (12_{10}); $D_{16} = 1101_2$ (13_{10}); $E_{16} = 1110_2$ (14_{10}); and $F_{16} = 1111_2$ (15_{10}). Hexadecimal is popular throughout computer systems, which often use 8-bit bytes or 16-bit or 32-bit words. These can conveniently be packaged as collections of 4-bit nibbles represented using the hexadecimal system. A count sequence illustrates the relationship between these different number systems:

The Assyrian radix was 60 and it still remains in our division of time. One plausible reason for the decimal system is that the number of our fingers and thumbs is ten. Don't infer too much about the Assyrian physiology!

Decimal	Binary	BCD	Octal	Hexadecimal
0	0	0000	0	0
1	1	0001	1	1
2	10	0010	2	2
3	11	0011	3	3
4	100	0100	4	4
5	101	0101	5	5
6	110	0110	6	6
7	111	0111	7	7
8	1000	1000	10	8
9	1001	1001	11	9
10	1010	00010000	12	A
11	1011	00010001	13	B
12	1100	00010010	14	C
13	1101	00010011	15	D
14	1110	00010100	16	E
15	1111	00010101	17	F
16	10000	00010110	20	10
17	10001	00010111	21	11
18	10010	00011000	22	12
19	10011	00011001	23	13
20	10100	00100000	24	14

This illustrates that a more compact number representation is achieved with a high radix, though this requires more symbols for the digits.

6.1.2 *Conversion between systems*

The most important code conversion step is from decimal to binary. This is because the other systems are essentially binary-based. In order to convert a decimal **integer** to binary, one algorithm is to divide a number by two and the remainder left at each stage becomes a bit in the converted binary number. The remainder from the first division is the least significant bit (LSB) of the resulting number and bits are generated in increasing order of significance. The number resulting from division is then divided by two and the remainder from division becomes the bit of next most significant precision. This process continues successively, until the result of division is one, the most significant bit (MSB) of the resulting number. Consider 9_{10} and 23_{10} converted into binary:

$9/2 = 4$ remainder 1 LSB $\qquad$ $23/2 = 11$ remainder 1 LSB
$4/2 = 2$ remainder 0 $\qquad\qquad$ $11/2 = 5$ remainder 1
$2/2 = 1$ remainder 0 $\qquad\qquad$ $5/2 = 2$ remainder 1
$1/2 = 0$ remainder 1 MSB $\qquad$ $2/2 = 1$ remainder 0
$\qquad\qquad\qquad\qquad\qquad\qquad\quad$ $1/2 = 0$ remainder 1 MSB

$9_{10} = 1001_2$ $\qquad\qquad\qquad\qquad$ $23_{10} = 10111_2$

In order to convert a binary integer to decimal, one method is to multiply the most significant bit by two and then add the bit of next least significance. This continues successively until the least significant bit is added. For 10111_2, which as just illustrated is 23_{10},

10111_2	10111_2
$1 \times 2 + 0$	$1 \times 2 + 0 = 2$
$(1 \times 2 + 0) \times 2 + 1$	$2 \times 2 + 1 = 5$
$((1 \times 2 + 0) \times 2 + 1) \times 2 + 1$	$5 \times 2 + 1 = 11$
$(((1 \times 2 + 0) \times 2 + 1) \times 2 + 1) + 1$	$11 \times 2 + 1 = 23_{10}$

and there are similar algorithmic techniques to convert **fractions**. Conversion between octal and binary and between hexadecimal and binary is much easier, since octal and hexadecimal are coded versions of binary. If we take a 12-bit word we can just partition it into three 4-bit sections for hexadecimal, or into four 3-bit sections for octal, e.g.

binary	0 1 0 0	1 1 0 1	0 1 0 1
octal	2	3 2	5
hexadecimal	4	D	5

6.1.3 Other codes

The MORSE code is of great importance historically because it was first used in telegraphic transmission. It is not used in digital circuits because the symbols vary in length. The *American Standard Code for Information Interchange* (ASCII) is a 7-bit alphanumeric code used in many computer systems and is the code used to represent characters, as is the *Extended Binary-Coded Decimal Interchange Code* (EBDIC). A code of major importance is the Gray code, which has already been introduced in ASM chart design (p. 114). This code is a minimal-distance code, since one bit only changes between elements of the code. It was originally used in rotational transducers to avoid the problems with alignment that a natural binary code would incur. *Excess 3* is a code that is equivalent to the BCD coding plus three, hence the name. This implies that the symbols use the binary codes for 3 to 12 inclusive and the codes for 0, 1, 2 and 13, 14, and 15 are not used.

Decimal	BCD	Excess 3
0	0000	0011
1	0001	0100
2	0010	0101
3	0011	0110
4	0100	0111

5	0101	1000
6	0110	1001
7	0111	1010
8	1000	1011
9	1001	1100

6.2   Binary arithmetic

6.2.1   Basic binary arithmetic

Basic arithmetic naturally concentrates on the main arithmetic functions, addition, subtraction, multiplication and division. We will consider **unsigned** numbers first and, to simplify presentation, we will consider positive integers only.

6.2.1.1   Binary addition

When adding two binary digits we generate a *sum* bit and a *carry* bit. The sum is the result of addition at that stage, and the carry is generated to be included in the addition stage of the bit of next most significance. The rules for binary, modulo 2, addition of two bits A and B, A + B, are

Digit 1	+	Digit 2	= Sum	Carry
0		0	0	0
0		1	1	0
1		0	1	0
1		1	0	1

When adding two binary numbers, say 110_2 (6_{10}) and 101_2 (5_{10}), these rules are used:

$$
\begin{array}{cccl}
1 & 1 & 0 & \\
1 & 0 & 1 & \\
\hline
1\ 0 & 1 & 1 & \text{sum} \\
1 & 0 & 0 & \text{carry} \\
\hline
1\ 0 & 1 & 1 & \text{result of addition}
\end{array}
$$

The result of adding these two numbers is then 1011_2, which is 11_{10} as expected. These rules can be implemented in **hardware** and we can design circuitry which implements binary addition. If the numbers are presented in parallel, then we need to design circuits that add two bits at a time and generate a carry to the next stage. The general format is:

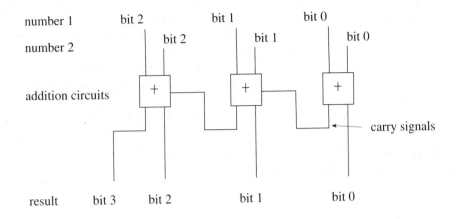

The addition rules provide a way to add two bits to provide a sum bit and a carry bit. The rules actually describe a truth table where the sum bit is '1' when either bit is '1' and is then the EXCLUSIVE OR of the two bits added. The carry bit is set only when both bits are '1' and is the AND of the two input bits. The circuit implementing the rules is known as a *half adder*.

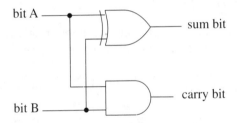

This circuit can be used to add the two least significant bits of a binary number. The rules for **full** addition describe the generation of a sum and carry bit for two input bits plus the carry generated from a previous stage. If the bits to be added are A_i and B_i, generating a sum bit S_i and a carry bit C_i, given the carry generated by the previous stage, the bit of next least significance is C_{i-1}. The rules for addition are then

C_{i-1} +	A_i +	B_i	S_i	C_i
0	0	0	0	0
0	0	1	1	0
0	1	0	1	0
0	1	1	0	1
1	0	0	1	0
1	0	1	0	1
1	1	0	0	1
1	1	1	1	1

The carry input at each stage is often known as the carry-in. The generated carry is often known as the carry-out, but don't confuse it with a pizza or a fourpack!

A *full adder* is a circuit which implements these rules. In order to design a minimised circuit for the sum bit S_i we can draw up its K-map:

C_{i-1} \ A_iB_i	00	01	11	10
0	0	1	0	1
1	1	0	1	0

The K-map does not give much help for minimisation in this case, since we can loop no common terms. By extracting each prime implicant alone we get

$$S_i = \overline{C}_{i-1} \cdot \overline{A}_i \cdot B_i + \overline{C}_{i-1} \cdot A_i \cdot \overline{B}_i + C_{i-1} \cdot \overline{A}_i \cdot \overline{B}_i + C_{i-1} \cdot A_i \cdot B_i$$

which by grouping terms can achieve an implementation using EXOR gates only

$$S_i = \overline{C}_{i-1} \cdot (\overline{A}_i \cdot B_i + A_i \cdot \overline{B}_i) + C_{i-1} \cdot (\overline{A}_i \cdot \overline{B}_i + A_i \cdot B_i)$$

$$= \overline{C}_{i-1} \cdot (A_i \oplus B_i) + C_{i-1} \cdot (\overline{A_i \oplus B_i})$$

$$= C_{i-1} \oplus A_i \oplus B_i$$

The carry bit C_i is '1' when two or more of the input bits are '1', so

$$C_i = C_{i-1} \cdot A_i + C_{i-1} \cdot B_i + A_i \cdot B_i$$

The circuit for a full adder is then

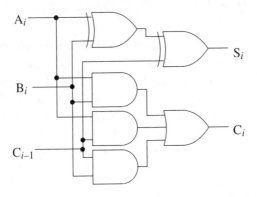

By further analysis of the carry signal we obtain

$$C_i = C_{i-1} \cdot (\overline{A}_i \cdot B_i + A_i \cdot \overline{B}_i + A_i \cdot B_i) + A_i \cdot B_i$$
$$= C_{i-1} \cdot (A_i \oplus B_i) + A_i \cdot B_i$$

and so a full adder can be implemented using two half adders and an OR gate.

By denoting a half adder as HA and a full adder as FA, the circuit to add two binary numbers is then as follows:

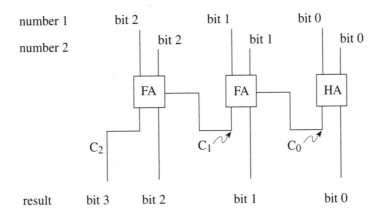

Arithmetic circuits have been a very fertile area of logic circuit development due to the importance of addition in computers. Carry look-ahead circuits can improve speed by testing the carry for all stages simultaneously, rather than the circuit provided where the carry propagates from one stage to the next, reducing potential speed.

6.2.1.2 *Binary subtraction*

When subtracting two binary digits we generate a *difference* bit which when '1' indicates that one bit is '1' and the other is '0' (hence indicating that there is a difference between them), and a *borrow* bit. The borrow is generated to indicate that the number subtracted is greater in size than the number it is being subtracted from (i.e. a negative result would ensue) and when subtracting bit B from bit A the borrow bit is '1' when A is '0' and B is '1'. Formally, the *minuend* is subtracted from the *subtrahend* to provide the difference between the two numbers. For two bits, to subtract B from A (to form $A - B$), the rules are

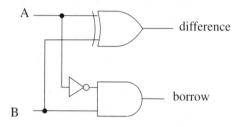

$A - B$		Difference	Borrow
0	0	0	0
0	1	1	1
1	0	1	0
1	1	0	0

Half subtraction rules   Half subtracter

These rules define a *half subtracter*, where the difference is identical to the sum bit for a half adder. The half subtracter is used for the LSBs and a *full subtracter* is used for the remainder of the bits, to subtract B_i from A_i given a borrow Br_{i-1} to generate a difference D_i and borrow Br_i

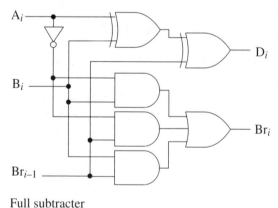

Br_{i-1}	A_i	B_i	D_i	Br_i
0	0	0	0	0
0	0	1	1	1
0	1	0	1	0
0	1	1	0	0
1	0	0	1	1
1	0	1	0	1
1	1	0	0	0
1	1	1	1	1

Full subtracter

Full subtraction rules   Full subtracter

We could then use the circuits for the full and half subtracters to provide a full circuit to subtract two binary numbers. This is usually avoided in signed binary arithmetic by using complement form, which employs addition only, and so we do not need subtracter circuits. We shall complete the basic binary arithmetic systems before looking at signed binary arithmetic.

6.2.1.3 Binary multiplication

In principle, binary multiplication is straightforward since anything multiplied by zero results in zero and anything multiplied by one remains the same. The rules for multiplication are then

A	× B	A×B
0	0	0
0	1	0
1	0	0
1	1	1

This is again a truth table and the AND function then implements a circuit to multiply two bits. The procedure for binary multiplication is similar to that often used for decimal multiplication. Formally, the number multiplied is called the *multiplicand*, the number multiplying it is called the *multiplier*.

In decimal arithmetic we introduce zeros corresponding to the position of the most significant digit and then multiply the multiplicand by that digit. On the next row we introduce zeros corresponding to the next most significant digit and then multiply by that digit. This continues to the row corresponding to the least significant digit, where we introduce no zeros and insert the result of multiplication. We then add up the results for each digit in to form the result. The results at each intermediate step are the *partial products*. For example, 53_{10} multiplied by 321_{10}

Multiplicand				5	3		
Multiplier			3	2	1	×	
(300×53)	1	5	9	0	0		Partial product for most significant digit 3
(20×53)		1	0	6	0		Partial product for 2
(1×53)				5	3	+	Partial product for least significant digit 1
	1	7	0	1	3	=	

The process of introducing zeros is equivalent to multiplying the multiplicand by a factor of ten, which, by virtue of the decimal system, is equivalent to shifting the number left and introducing zeros into the vacated digits, and this can be seen in the partial products. The same procedure is used in binary multiplication, the multiplicand is shifted left according to the position or significance of the bit in the multiplier. The results are all summed to provide the final result, and so the process is called *shift and add*. Each partial product is formed by multiplying a binary number by a factor of two, which is equivalent to shifting the number left and adding zeros to the vacated positions. For multiplication by 4, we shift the number left two places, e.g. $1011_2 \times 100_2$ (2^2).

$$1 \quad 0 \quad 1 \quad \times 1 \quad 1 \quad 0 \quad 0 \quad = \quad 1 \quad 0 \quad 1$$

This is used to form each partial product when multiplying by a binary integer, e.g. 1011_2 multiplied by 1101_2:

Multiplicand				1	0	1	1
Multiplier				1	1	0	1	$\times$

(1000×1011)	1	0	1	1	0	0	0	
(100×1011)		1	0	1	1	0	0	
(1×1011)				1	0	1	1	$+$

1	0	0	0	1	1	1	1	$=$

The resulting binary number is then 143_{10} (the input numbers are 13_{10} and 11_{10}). One partial product is omitted in the summation. This is because one bit is 0 in the multiplier (corresponding to 2^1).

The shift and add technique lends itself readily to a hardware implementation. Since multiplying by 2^n is equivalent to shifting left by n places, multiplication can then be implemented using a shift register. The outputs of the shift register are gated from another shift register holding the multiplier to give an input to an adder circuit. The multiplier is stored in a shift register to facilitate testing whether bits are set or not. The other number input to the adder is from the outputs of a register, which finally holds the result of multiplying the two numbers, formed by accumulating the partial products.

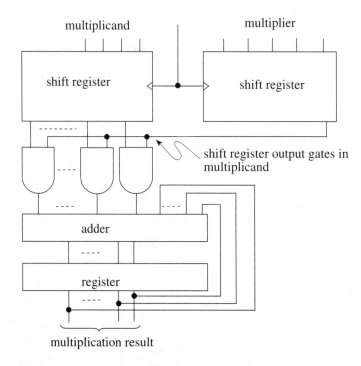

At the first clock cycle the multiplicand is fed to the adder only if the multiplier LSB is '1'; otherwise the first partial product is zero. The first partial product is then stored in the register. At the next clock cycle the multiplicand is shifted left one place (multiplying it by 2), and if the next most significant bit in the multiplier is '1' then this second partial product is added to the first. The register then contains the result of adding the first two partial products. Multiplication continues in this manner by shifting the multiplicand to form the partial products and adding it to the contents of the register according to whether appropriate bits are set in the multiplier. The process terminates when the partial product resulting from the most significant bit is added to the final sum. The register then contains the result of multiplication. The register needs to be sufficiently large to accommodate the result, and in this case the size should at least be the joint length of the multiplier and the multiplicand. In a software version, the procedure to multiply `multiplicand` by `multiplier` is:

```
FOR i=1 TO number_of_bits DO
  BEGIN
  IF (multiplier[i]='1') result:= result (plus) multiplicand
  ;If the bit is set, add the partial product to the result
  multiplicand=multiplicand×2
  ;Now shift the multiplicand left to form the next partial product
  END
```

This is not a PALASM specification and is not directly suited to a PALASM implementation, as PALASM does not support the counting loop, `FOR i=1 TO N` (except in simulation), or arithmetic addition. Multiplication of binary integers is easily accomplished using a look-up table, the implementation is discussed in Section 6.3.1.

6.2.1.4   Binary division

Binary division is more complex. It can use a procedure equivalent to that for conventional longhand division. Formally, the number that is divided is known as the *dividend*, the number it is divided by is the *divisor*, and the result is the *quotient*. Note that we can form a result with a fractional part. At each stage, we determine how many times the divisor is an integer number of divisions of the dividend. We start with the most significant number of divisions and work towards the bits of least significance. At each stage we perform binary subtraction and bring down the next least significant digit from the dividend. We can add trailing zeros to the dividend to increase the number of digits in a fractional result. Take, for example, 1110_2 divided by 100_2:

```
          1  1. 1                        3. 5
100 | 1  1  1  0. 0                  4 | 14. 0
      1  0  0                            12
         1  1  0                          2 0
         1  0  0                          2 0
            1  0  0                        0 0
            1  0  0
            0  0  0
```

result of dividing 14_{10} by 4_{10} is then 3.5_{10}, as expected, and in this case the procedure
minated with an exact fraction. This is not usually the case where division continues until
specified number of bits have been generated. This procedure is highly complex to
implement as a digital system.

> Note that you can store the representation of the
> reciprocal of a number in memory. Division then
> becomes a process of multiplication by the
> reciprocal of a number, rather than dividing by the
> number, and this is much faster.

6.2.2   Signed binary arithmetic

6.2.2.1   Signed arithmetic

In order to handle signed binary numbers we could use an extra bit, a *sign bit*, and set it to
'0' for positive numbers or '1' for a negative number. When adding or subtracting two
numbers we could use the sign bit to choose the output from an adder or a subtracter
respectively. This implies that we need different circuits to add and subtract numbers; and,
as shown earlier, though the sum bit is the same as the difference, the borrow is completely
different from the carry. Using separate adders and subtracters would therefore increase
cost. It is better to use *complement arithmetic*, where we represent a number in its
complement, or negative form, and use addition only. Instead of performing a subtraction
A – B, we form the complement of B, –B, and add this to A to form the subtraction as

$$A - B = A + (-B)$$

Though we then need the means to generate the complement form of a number, we only
need addition units and not those for subtraction as well. This is much cheaper than using
adders and subtracters in tandem.

There are two ways to generate the complement: there is a **radix** complement and a
radix – 1 complement. In decimal arithmetic there is a *ten's complement* and a *nine's
complement*; in binary there is a *two's complement* and a *one's complement* for the radix and
radix – 1 complement respectively. For an *n*-digit number N, the radix complement is
$r^{n+1} - N$ and the radix – 1 complement is $(r^{n+1} - 1) - N$. For an *n*-digit number N,

Ten's complement $10^{n+1} - N = 10000...00 - N$
Nine's complement $10^{n+1} - 1 - N = 10000...00 - 1 - N = 9999...99 - N$
Two's complement $2^{n+1} - N = 1000...00 - N$
One's complement $2^{n+1} - 1 - N = 1000...00 - 1 - N = 111...11 - N$

When the number is in complement form we have no way of knowing this. We therefore
include a designated sign bit that in binary will be '1' to indicate a negative number and '0'
a positive one. Note that we include the sign bit as '0' prior to complementing a number.

When the number is in complement form the sign bit will be '1'. Having formed the complement, we then need to perform addition only and have no need for extra subtraction circuitry.

Note that excess 3 is sometimes termed self-complementing owing to the cyclic nature of the complements of elements of the code. The nine's complement of 3 is 6 and that of 9 is 1.

As well as including an extra bit to indicate the sign of a number, we can include a further *overflow* bit to handle a situation where the result might be too large to fit within a specified bit pattern. For example, if we add two 4-bit numbers 1000_2 (8_{10}) and 1001_2 (9_{10}), we generate a result 10001_2 (17_{10}), which is a 5-bit number with the fifth bit generated by the size of the two 4-bit numbers. If we did not specify an extra bit to accommodate the overflow then this number could be interpreted to be a negative 4-bit number (since the sign bit is '1'), a confusing result. To add these numbers, the fifth bit is the overflow bit and the sixth bit is the sign bit. The overflow bit will not be included in examples here but needs to be borne in mind by a designer.

6.2.2.2   One's complement arithmetic

In order to generate the one's complement we need to subtract the number from a set of '1's. For a 3-bit number $B_2B_1B_0$, we first include an extra bit as the sign bit, the (positive) number then becomes $0B_2B_1B_0$, and we then subtract the number from a set of '1's to form the complement

$$
\begin{array}{cccc}
1 & 1 & 1 & 1 \\
- 0 & B_2 & B_1 & B_0
\end{array} \quad \text{(positive) original number}
$$

$$
= 1 \quad \overline{B_2}\,\overline{B_1}\,\overline{B_0} \quad \text{(negative) number in one's complement form}
$$

This shows that **to form the one's complement** we merely need to **invert each bit**. The sign bit becomes '1', indicating that the number is in complement form. We can now add to perform subtraction. For example, $2_{10} - 8_{10}$, where $2_{10} = 0010_2$ and $8_{10} = 1000_2$. We first add a sign bit to each number, then complement 8_{10} and then add the complemented form of 8_{10} to 2_{10}:

8_{10} 1000   plus sign bit 01000   one's complement 10111

To perform $2_{10} - 8_{10}$:   +2   00010
                                  −8   10111

                     Addition   11001

By inspecting the result we can see that the sign bit is '1'. This implies that the resulting number is negative and in one's complement form. To find out what the number is we merely need to invert each bit to see that it represents -00110 or -6, the expected result.

The number has only been inverted here to illustrate that the correct result has been achieved. When complement arithmetic is used in arithmetic in computers, or in other digital systems, the complement is generated only at the input or at the output stage. At the input stage we can feed negative numbers to a digital system by generating their complement, and at the output stage we can invert the complement to see the number the complement represents. For the remainder of the arithmetic process the number remains in its complement form; we do not need to generate complements during arithmetic, only before or after.

Consider also $8_{10} - 2_{10}$. Given that $8_{10} = 1000_2$ and $2_{10} = 0010_2$. We need to add a sign bit to each number and complement 2_{10}, then add the complemented form to 8_{10}:

2_{10} $\qquad$ 0010 $\quad$ plus sign bit 00010 $\quad$ one's complement 11101

To perform $8_{10} - 2_{10}$: $\qquad$ +8 $\qquad$ 01000
$\qquad\qquad\qquad\qquad\qquad$ −2 $\qquad$ 11101

$\qquad\qquad\qquad$ Addition 100101

When we inspect the sign bit we can see that it represents a positive number, since the sign bit, bit 4, is '0'. Our main problem is that there is an extra bit beyond the sign bit and the other difficulty is that the answer appears to be wrong: − the result of $8 - 2 = 5$! This is actually inherent in one's complement arithmetic since we have performed

$$8 + 2^{n+1} - 1 - 2 = 5 + 2^{n+1},$$

the given result. All the extra bit tells us is that the number is out by one by virtue of one's complement arithmetic. We then add the extra '1' (if it is generated) back into the LSB in a process known as *end-around-carry*; e.g. $14_{10} - 3_{10}$:

$\qquad\qquad\qquad\qquad$ 3 $\qquad\qquad\qquad\qquad\qquad\qquad$ 00011
$\qquad\qquad\qquad\qquad$ −3 $\qquad\qquad\qquad\qquad\qquad\qquad$ 11100
$\qquad\qquad\qquad\qquad$ 14 $\qquad\qquad\qquad\qquad\qquad\qquad$ 01110
$\qquad\qquad\qquad\qquad$ 14 − 3 $\qquad\qquad\qquad\qquad\qquad$ 101010
$\qquad\qquad\qquad\qquad$ end-around-carry $\qquad\qquad\qquad\qquad\quad$ 1+
$\qquad\qquad\qquad\qquad$ result $\qquad\qquad\qquad\qquad\qquad\qquad$ 01011=

When the end-around-carry is implemented, the result is 11_{10}, as expected. Since the end-around-carry is inherent in one's complement arithmetic, it makes it a **two-stage process**. There are two addition stages; one is to correct the result. By virtue of this, two's complement arithmetic is universally preferred, since it does not need this extra correction stage. Also, there are two representations of zero in one's complement arithmetic; both 0000 ... 0

and 1111 ... 1 represent zero (the latter in one's complement form), and this can lead to extra difficulty.

6.2.2.3   Two's complement arithmetic

The two's complement of an n-bit number N is defined as $2^{n+1} - N$. This is the one's complement plus one. The one's complement was generated by inverting each digit, so **to form the two's complement** we need to **invert each bit then add one**. We need to include an extra sign bit and after the complementing process the sign bit will be '1', indicating a negative number in two's complement form. Again, having formed the complement we then need to perform addition only and we do not need subtraction; e.g. $2_{10} - 8_{10}$:

$$8_{10} = 1000 \text{ add sign bit, } +8_{10} = 01000 \text{ two's complement} = \text{invert each bit and add one}$$
$$- 8_{10} \qquad\qquad = 10111 + 1$$
$$= 11000$$

$$\begin{array}{rcl} +2_{10} = & 00010 \\ -8_{10} = & 11000 \\ \hline 2 + (-8) = & 11010 \end{array}$$

When we inspect the result, the sign bit is '1', indicating a negative number in two's complement form. To determine the negative number it represents we merely repeat the complementing process, invert each bit and add one:

$$11010 = -00101 + 1 = -00110 = -6_{10}$$

This shows that the result is as expected. Note again that once the number is in complement form, it remains so for the remainder of the arithmetic process. As with one's complement arithmetic we are only inverting the complemented number to show that the correct result has been achieved.

Finally, let us consider $8_{10} - 2_{10}$:

$$\begin{array}{rcl} +2_{10} & = & 00010 \\ -2_{10} & = & 11101 + 1 \\ & = & 11110 \\ +8_{10} & = & 01000 + \\ \hline & & 100110 = \end{array}$$

If we ignore the extra bit above the sign bit, the result is 00110, $+6_{10}$ as expected. The extra bit can actually be ignored since it is inherent in two's complement arithmetic. We have performed $8 + 2^{n+1} - 2 = 6 + 2^{n+1}$. The extra bit corresponds to 2^{n+1}, which was only introduced as part of the complementing process and can therefore be ignored. Two's complement arithmetic has **one stage only**, as opposed to one's complement, which requires two stages. For this reason, two's complement arithmetic is universally preferred. Also, there is a single representation of zero, 0000 ... 0, in two's complement arithmetic by virtue of the complementing process.

6.2.3 Fixed- and floating-point arithmetic

General-purpose arithmetic systems need to handle numbers with both fractional and integer parts. Two representations exist, called *floating-point* and *fixed-point* arithmetic. Floating-point arithmetic represents a number as a *mantissa* and *exponent*. The mantissa is a binary fraction expressed in two's complement form. The exponent is binary and again expressed in complement form. The floating-point representation is then

$$\text{number} = \text{mantissa} \times \text{exponent} = m \times 2^e$$

where m, the mantissa, is a fractional binary number, $m = \sum d_{-i} 2^{-i}$ where $i \in 1, n$ for an n-bit mantissa, and e is the exponent. This is expressed as a pattern of bits, usually stored within a computer word. Floating-point systems usually employ a specified format, among which the IEEE/ANSI 754-1985 is often preferred in computer systems. Fixed-point systems are usually application-specific for reasons of speed, and use a floating-point number system arranged explicitly for the application and generally without using an exponent as in floating-point systems.

An arrangement of a word for a floating-point number is then

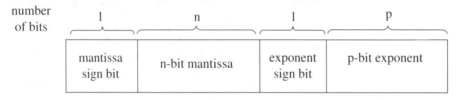

The mantissa is usually normalised so that $0.5 \leq \text{mantissa} < 1.0$. This is consistent with the binary exponent, and since the first bit of the number is 1 it ensures that all the available *resolution* is used. The resolution describes how well the binary number approximates its true version. The resolution is limited, since we are only storing a limited, and not infinite, set of bits to represent a number. Common arithmetic systems in computers use 16-bit and 32-bit arithmetic, and some languages allow the use of extra *precision* to store numbers to expedite better accuracy. For example, in a system with a 6-bit mantissa and a 3-bit exponent,

$$1.6875_{10} = 0.84375 \times 2^1$$
$$= (1 \times 0.5 + 1 \times 0.25 + 0 \times 0.125 + 1 \times 0.0625 + 1 \times 0.03125) \times 2^1 = 0.11011/001$$

The normalisation process reduces the number to be within the correct range and the exponent is increased to reflect by how much the mantissa has been reduced. The slash between the two binary numbers indicates the partition between the mantissa and exponent.

If the normalised number has been increased to fit within the range then the exponent will be negative, indicating by how much the mantissa needs to be reduced to return to the original number. The negative exponent will be stored in two's complement form; for example, 0.3128_{10} needs to be multiplied by 4 to fit within the normalisation range, so

$$0.1328_{10} = 0.53125 \times 2^{-2}$$
$$= (1 \times 0.5 + 0 \times 0.25 + 0 \times 0.125 + 0 \times 0.0625 + 1 \times 0.03125) \times 2^{-2} = 0.1001/110$$

where the leading '1' in the exponent is a sign bit indicating a number expressed in two's complement form representing -2. The size of the exponent restricts it to between -4 (100) and $+3$ (011).

In order to add two floating-point numbers they first need to be denormalised so that the exponents are equal. The mantissas are then of the same range and they can be added. The number with the smallest exponent is denormalised by dividing it by two and then increasing the exponent by one. This double process continues until the exponent equals that of the other number, when the two mantissas represent fractions in the same range. For example, $1.6875_{10} + 0.1328_{10}$

$1.6875_{10} = $ 0.11011/001;this has the largest of the two exponents and is thus the target for normalisation

$0.1328_{10} = $ 0.10001/110, so shift the mantissa right and add one to the exponent
$= $ 0.01000/111; the exponent is now -1, so continue
$= $ 0.00100/000; the exponent is now zero, so we need one more step
$= $ 0.00010/001 the final denormalised number

	1.6875_{10}		0.11011/001
+	0.1328_{10}	+	0.00010/001
=	1.8203_{10}	=	0.11101/001

The result of binary floating-point addition actually represents 1.8125_{10}, which does not equal the result of decimal arithmetic owing to the decreased resolution of the 6-bit mantissa in the floating-point representation. Information was lost during denormalisation of 0.1328_{10} (the final '1' in the normalised number was lost during denormalisation, since it did not fit within a 6-bit mantissa). This is the effect of limited precision arithmetic.

It is possible to implement floating-point arithmetic in hardware to improve speed in many modern systems. This is often implemented using a floating-point coprocessor alongside a computer's central processing unit.

The processes of multiplication and division are given mathematically as follows:

multiplication $\quad (A \times 2^a) \times (B \times 2^b) = (A \times B) \times 2^{(a+b)}$

division $\qquad\quad (A \times 2^a) / (B \times 2^b) = (A / B) \times 2^{(a-b)}$

These processes might require renormalisation of the resulting mantissa to ensure that it remains within the normalisation range.

6.3 Design examples

A number of design examples are included to illustrate how arithmetic systems not only can be implemented, in particular within programmable combinational circuits, but also can be used for design purposes to facilitate implementation. Arithmetic circuits are of natural interest to computer designers and there has been much research directed at their implementation. The presentation here is to clarify the main academic material while providing pointers to the issues that system designers need to resolve.

6.3.1 Binary arithmetic

We shall first consider the design of a circuit to multiply two 2-bit positive binary integers. This can be implemented as a combinational logic design, since the product of the two numbers is fixed. The combinational logic design can be specified using a truth table with a design based on a K-map. The K-map is a rather prolix approach, though it is good exercise in combinational design. To multiply a binary number A_1A_0 by B_1B_0 to produce a result $C_3C_2C_1C_0$ where A_1, B_1 and C_3 are the MSBs, the truth table is

A_1A_0	$\times$	B_1B_0	$C_3C_2C_1C_0$		A_1A_0	$\times$	B_1B_0	$C_3C_2C_1C_0$
0 0		X X	0 0 0 0		1 0		0 1	0 0 1 0
X X		0 0	0 0 0 0		1 0		1 0	0 1 0 0
0 1		0 1	0 0 0 1		1 0		1 1	0 1 1 0
0 1		1 0	0 0 1 0		1 1		0 1	0 0 1 1
0 1		1 1	0 0 1 1		1 1		1 0	0 1 1 0
					1 1		1 1	1 0 0 1

By K-map analysis,

$$C_0 = A_0 \cdot B_0$$

$$C_1 = A_1 \cdot \overline{B}_1 \cdot B_0 + A_1 \cdot \overline{A}_0 \cdot B_0 + A_0 \cdot \overline{B}_0 \cdot B_1 + \overline{A}_1 \cdot A_0 \cdot B_1$$

$$C_2 = A_1 \cdot B_1 \cdot \overline{B}_0 + A_1 \cdot \overline{A}_0 \cdot B_1$$

$$C_3 = A_1 \cdot A_0 \cdot B_1 \cdot B_0$$

A simpler method of designing the circuit is to use shift-and-add as the design process. The multiplication result is the sum of $B_1 B_0$ shifted right one place, which is added to a sum if A_1 is set. $B_1 B_0$ is added to the sum if A_0 is set. The design can be achieved by evaluating the sum bit at each stage, given the carry generated by the previous stage:

$$
\begin{array}{cccc}
 & B_1 & B_0 & \\
 & A_1 & A_0 & \times \\
\hline
B_1{\cdot}A_1 & B_0{\cdot}A_1 & 0 & \\
 & B_1{\cdot}A_0 & B_0{\cdot}A_0 & + \\
\hline
C_3 \ \ C_2 & C_1 & C_0 & = \\
\end{array}
$$

$C_0 = 0$ plus $B_0{\cdot}A_0 = B_0{\cdot}A_0$

This generates no carry.

$C_1 = B_0{\cdot}A_1$ plus $B_1{\cdot}A_0$ given carry of $0 = B_0{\cdot}A_1 \oplus B_1{\cdot}A_0$

$\qquad\qquad$ by de Morgan's law $= B_0{\cdot}A_1{\cdot}(\overline{B}_1 + \overline{A}_0) + (\overline{B}_0 + \overline{A}_1){\cdot}B_1{\cdot}A_0$

This generates a carry $= B_0{\cdot}A_1{\cdot}B_1{\cdot}A_0$.

$C_2 = B_1{\cdot}A_1$ plus the carry from the previous stage $= B_1{\cdot}A_1 \oplus B_0{\cdot}A_1{\cdot}B_1{\cdot}A_0$

$\qquad\qquad = (\overline{B}_1 + \overline{A}_1){\cdot}B_0{\cdot}A_1{\cdot}B_1{\cdot}A_0 + B_1{\cdot}A_1{\cdot}(\overline{B}_0 + \overline{A}_1 + \overline{B}_1 + \overline{A}_0)$

$\qquad\qquad = B_1{\cdot}A_1{\cdot}\overline{B}_0 + B_1{\cdot}A_1{\cdot}\overline{A}_0$

This generates a carry $= B_1{\cdot}A_1{\cdot}B_0{\cdot}A_1{\cdot}B_1{\cdot}A_0 = B_1{\cdot}A_1{\cdot}B_0{\cdot}A_0$

$C_3 =$ the carry from the previous stage $\qquad = B_1{\cdot}A_1{\cdot}B_0{\cdot}A_0$

These equations are the same as those by combinational design, merely derived by mathematical analysis rather than by K-maps, illustrating the duality between the two techniques.

6.3.2 Code conversion

6.3.2.1 Binary to BCD conversion

We will now design a system to convert from 4-bit binary to BCD. The conversion we want to achieve is

Binary $B_3B_2B_1B_0$	BCD $D_4D_3D_2D_1D_0$	Binary $B_3B_2B_1B_0$	BCD $D_4D_3D_2D_1D_0$
0 0 0 0	0 0 0 0 0	1 0 0 0	0 1 0 0 0
0 0 0 1	0 0 0 0 1	1 0 0 1	0 1 0 0 1
0 0 1 0	0 0 0 1 0	1 0 1 0	1 0 0 0 0
0 0 1 1	0 0 0 1 1	1 0 1 1	1 0 0 0 1
0 1 0 0	0 0 1 0 0	1 1 0 0	1 0 0 1 0
0 1 0 1	0 0 1 0 1	1 1 0 1	1 0 0 1 1
0 1 1 0	0 0 1 1 0	1 1 1 0	1 0 1 0 0
0 1 1 1	0 0 1 1 1	1 1 1 1	1 0 1 0 1

We wish to design a system which, given the 4-bit binary input, will produce the appropriate 5-bit binary output. There are a number of ways to achieve the design, and we could use

    (a) memory;
    (b) combinational design via K-maps;
    (c) combinational design via mathematics.

The memory implementation would require specifying the required conversion table as the contents of a ROM. Each of the five outputs has sixteen slots addressed by the four input bits. This requires a ROM with eighty locations only. This is a comparatively expensive solution. Cost can be reduced either by using a PAL or, in a chip design, by reducing the gate count. The design is therefore performed first by traditional K-map design, which is compared with the result of a PAL implementation. K-maps are really limited to small-scale problems only, so we will also use two's complement arithmetic for minimisation. This technique is much better suited to circuits with more inputs, which can be described using mathematical techniques.

The minimisation techniques should all achieve the same (or an equivalent) result, which they do. They reach this solution by different ways, each highlighting the benefits and practice of a chosen design technique.

6.3.2.2 Combinational design

By inspection of the code conversion to be implemented we can see that D_0 is a direct copy of B_0. So the output D_0 is given by $D_0 = B_0$. By K-map analysis for D_1 and D_2,

$$D_1 = \overline{B_3} \cdot B_1 + B_3 \cdot B_2 \cdot \overline{B_1}$$

$$\overline{D_2} = \overline{B_3} \cdot B_2 + B_2 \cdot B_1$$

D_3 is set only in two cases, so minimisation does not need a K-map. D_4 is '1' only when B_3 is '1' and B_2 or B_1 is '1'.

$$D_3 = B_3 \cdot \overline{B_2} \cdot \overline{B_1}$$

$$D_4 = B_3 \cdot (B_2 + B_1)$$

The specification for a PAL can just describe the BCD outputs using a CASE statement constructed from the vectored binary inputs:

```
CHIP binbcd PALCE16V8
;binary to BCD converter

PIN   5..8    bin[0..3]   COMBINATORIAL;4-bit binary input
PIN   14..18  bcd[0..4]   COMBINATORIAL;5-bit bcd output
PIN   10      GND
PIN   20      VCC

EQUATIONS

CASE (bin[3..0])   ;CASE statement
BEGIN        ;implementing look-up table
   10: BEGIN bcd[4..0] = #b10000 END
   11: BEGIN bcd[4..0] = #b10001 END
   12: BEGIN bcd[4..0] = #b10010 END
   13: BEGIN bcd[4..0] = #b10011 END
   14: BEGIN bcd[4..0] = #b10100 END
   15: BEGIN bcd[4..0] = #b10101 END
```

```
OTHERWISE: BEGIN
           bcd[4]=0 ;set MSB to 0
           bcd[3..0] = bin[3..0]
           END  ;and copy across binary value
END
```

which when compiled achieves

```
BCD[4]  =  BIN[1] * BIN[3]  +  BIN[2] * BIN[3]
BCD[3]  =  BIN[3] * /BIN[1] * /BIN[2]
/BCD[2] =  /BIN[2]  +  /BIN[1] * BIN[3]
BCD[1]  =  BIN[1] * /BIN[3]  +  /BIN[1] * BIN[2] * BIN[3]
/BCD[0] =  /BIN[0]
```

This is the same result as by K-map extraction.

6.3.2.3 *Minimisation by mathematics*

An alternative method for designing the binary-to-BCD converter can be derived from examining the required functionality. If we inspect the BCD code it is the same as the binary input (though with an extra '0' as the MSB) for the first nine codes from 0000_2 up to and including 1001_2. When the binary number is ten or more then the output is the input minus ten, with the extra bit set to '1'. This can be summarised in software by denoting the 4-bit binary input as input[3..0] and the 5-bit BCD output as output[4..0]. The MSBs are input[3] and output[4].

```
IF (input>=0 AND input<=9) THEN BEGIN ;copy input to the output
                                output[4] = 0 ;MSB = 0
                                output[3..0] = input[3..0]
                                END
                           ELSE BEGIN ;subtract ten from input
                                output[4] = 1 ;MSB =1
                                output[3..0] = input[3..0] - ten
                                END
```

Again, this does not lend itself to a direct PALASM description, since there is no arithmetic support; it can be done in PALASM but is tedious. It does, however, lead to a possible hardware implementation using an adder to subtract ten from the number, a multiplexer to choose either a copy of the input or the input with ten subtracted from it, and a circuit to generate the control signal for the multiplexer:

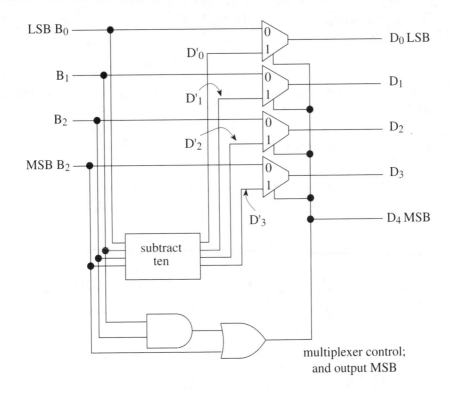

multiplexer control;
and output MSB

This gives the possibility for using chips in an MSI solution, since the 74157 is a quad 2-1 multiplexer and the 74283 is a 4-bit adder circuit. The control signal for the multiplexer is '1', to indicate when the input number is between ten and fifteen, and '0' otherwise. This is actually the signal $D_4 = B_3 \cdot (B_2 + B_1)$ in the combinational design, since this signal is '1' and '0' for the appropriate input ranges. This gives a control signal to the multiplexer, choosing the four input bits direct when '0', and the outputs of the adder when '1'. D_4 is also the MSB of the 5-bit BCD output. The MSI solution is then rather profligate, since it needs three chips at least, whereas we have already seen that the whole design can fit within a single PAL. The fact that the LSB of the BCD output is identical to the LSB of the binary input (as shown in the previous section) reflects this. A single PAL confers the standard advantages of a programmable implementation. It is actually possible to reduce the adder/multiplexer circuit to equal that derived by combinational logic design. This is achieved by using complement arithmetic for minimisation that is actually suited to the algebraic minimisation of circuits that can be expressed mathematically.

The adder circuit adds -1010_2 to the binary input; it can in principle be reduced, since we are subtracting a constant number using a generalised adder (which works for any number). It can be minimised by using either binary subtraction or complement arithmetic. We shall use two's complement arithmetic here, though the results are the same for both techniques. One's complement is much more complex, since we need to handle the end-around-carry. The two's complement of a number is given by inverting each bit and then adding one. In order to subtract 10_{10} from the binary input we need to perform

$$\begin{array}{ccccccccccccc}
B_3 & B_2 & B_1 & B_0 & = & B_3 & B_2 & B_1 & B_0 & & = & B_3 & B_2 & B_1 & B_0 \\
-1 & 0 & 1 & 0 & + & 0 & 1 & 0 & 1 & +1 & & +0 & 1 & 1 & 0 \\
& & & & & & & & & & = & D'_3 & D'_2 & D'_1 & D'_0
\end{array}$$

D'_0 is given by the addition of B_0 plus 0. Summation is the EXOR function, and with one input '0' the EXOR gate acts as a buffer:

$$D'_0 = B_0, \text{ with a carry} = \text{'0'}$$

D'_1 is given by B_1 plus 1 with the carry from the previous stage. This carry is '0' and so the summation is $\overline{B_1}$ since the EXOR gate acts as an inverter when one input is '1':

$$D'_1 = \overline{B_1}, \text{ with a carry} = B_1$$

D'_2 is given by B_2 plus 1 with the carry from the D'_1. This carry is B_1 and so the summation is B_1 EXOR $\overline{B_2}$.

$$D'_2 = B_1 \cdot B_2 + \overline{B_1} \cdot \overline{B_2}, \text{ with a carry} = B_1 + B_2$$

D'_3 is given by B_3 plus 0 with the carry from the D'_2. This carry is generated when either B_2 or B_1 is '1', so this carry is $B_1 + B_2$ and so the summation is then $(B_1 + B_2)$ EXOR B_3.

$$D'_3 = (B_1 + B_2) \oplus B_3, \text{ generating a carry} = (B_1 + B_2) \cdot B_3$$

We can ignore the sign bit in the result since we only use the output of the subtracter when the binary input is between 1010_2 and 1111_2, when the subtracter gives a positive output. We can then ignore the carry generated by D'_3.

This gives a minimised circuit for the adder. It can be minimised only because we are subtracting a constant, specified, number. We can also minimise the whole system. The multiplexer gates through either the binary input or the output from the subtracter. The LSB is the same in both cases so one of the 2-1 multiplexers is not needed at all. The multiplexer outputs are D_3, D_2, D_1 and D_0 chosen by a control signal. The control signal is $B_3 (B_2 + B_1)$, which when '0', selects the binary input, and when '1' selects the subtracter output.

For the LSB

$$D_0 = \overline{\text{control}} \cdot B_0 + \text{control} \cdot D'_0 \text{ where } D'_0 = B_0$$

so

$$D_0 = \overline{\text{control}} \cdot B_0 + \text{control} \cdot B_0 = B_0$$

For D_1,

$$D_1 = \overline{control} \cdot B_1 + control \cdot D'_1$$

$$= \overline{control} \cdot B_1 + control \cdot \overline{B}_1$$

$$= \overline{B_3 \cdot (B_2 + B_1)} \cdot B_1 + B_3 \cdot (B_2 + B_1) \cdot \overline{B}_1$$

by de Morgan's law

$$= (\overline{B}_3 + \overline{B}_2 \cdot \overline{B}_1) \cdot B_1 + B_3 \cdot B_2 \cdot \overline{B}_1$$

$$= \overline{B}_3 \cdot B_1 + B_3 \cdot B_2 \cdot \overline{B}_1$$

For D_2,

$$D_2 = \overline{control} \cdot B_2 + control \cdot D'_2$$

$$= \overline{control} \cdot B_2 + control \cdot (B_1 B_2 + \overline{B}_1 \cdot \overline{B}_2)$$

$$= \overline{B_3 \cdot (B_2 + B_1)} \cdot B_2 + B_3 \cdot (B_2 + B_1) \cdot (B_1 \cdot B_2 + \overline{B}_1 \cdot \overline{B}_2)$$

by de Morgan's law

$$= (\overline{B}_3 + \overline{B}_2 \cdot \overline{B}_1) \cdot B_2 + B_3 \cdot B_2 \cdot B_1 = \overline{B}_3 \cdot B_2 + B_3 \cdot B_2 \cdot B_1$$

$$= \overline{B}_3 \cdot B_2 + B_2 \cdot B_1$$

For D_3

$$D_3 = \overline{B_3 \cdot (B_2 + B_1)} \cdot B_3 + B_3 \cdot (B_2 + B_1) \cdot (B_3 + (\overline{B}_1 + B_2))$$

which reduces to

$$D_3 = B_3 \cdot \overline{B}_2 \cdot \overline{B}_1$$

These reduced equations are identical to the earlier combinational design. This shows not only that there is redundancy within the MSI solution, but also how mathematics can be used to reduce circuits.

6.3.3 Adder/magnitude comparator

We shall now design a circuit to compare the magnitude of two binary numbers to provide signals indicating that the input is greater than, less than, or equal to a threshold input. This requires a more sophisticated design than those presented earlier. The design uses a full adder with two's complement addition to subtract the two numbers. If the resulting sign bit is '0', then the result of addition is positive, whereas if the sign bit is '1' then the result is negative and the input must be less than the threshold. The adder/magnitude comparator subtracts a 3-bit threshold from a 3-bit input. The PALASM language does not include mathematical functions, so the full adder is expressed using logical equations. The threshold is expressed in two's complement form, so the expression for the sum bit for the LSB is the LSB of the input plus the inverted LSB of the threshold plus one, which is consistent with two's complement arithmetic. This reduces to the EXOR function of the input LSB and the threshold LSB. The carry generated by this addition reduces to the input LSB or the inverted threshold LSB. All other stages employ the full adder equations (Section 6.2.2.1), given that the threshold bits are inverted. The design accepts both the input and the threshold as 3-bit two's complement numbers. In order to handle overflow an extra bit is included within the generated sum. The generated sum is then 4-bit, with the MSB as the sign bit. In the addition, the extra overflow bit introduced into the summation result equates to the introduction of a fourth bit, into both the input and the threshold, equal to the sign bit in the presented numbers. By virtue of two's complement arithmetic this is equivalent to requiring the device to subtract two 4-bit numbers, but without including overflow.

Implementation of a full adder requires sum and addition circuitry. This leads to an explosion in the number of product terms, since the sum bit is the EXOR of three inputs, which is four prime implicants and hence four product terms. If one of these inputs is actually more complex than a single input, as in a carry generated by a previous stage, then the number of product terms becomes too large to fit within a PAL. This can be handled using intermediate output stages, labelling one of the PAL outputs as a logic function of specified inputs, to break up a term requiring too many product terms into one that will fit into the product terms for two or more PAL outputs. For the adder/magnitude comparator the carry signal generated by each stage is labelled as an intermediate output. Even though this reduces the available number of outputs, it means that the design fits within the PAL. We then require three bits to handle the carry signals together with the four bits for the summation result. There is one signal left available as an output, and this is used for a signal indicating that the presented numbers were equal. Ideally we would have separate signals for greater than and less than, but this is not possible within the confines of a PALCE16V8. The design is then

```
CHIP     magcomp  PALCE16V8
;adder/magnitude comparator
;input and threshold are both 3-bit two's complement numbers

PIN 5..7    input[2..0]     COMBINATORIAL ;3-bit binary input
PIN 2..4    threshold[2..0] COMBINATORIAL ;3-bit threshold
PIN 11      oe              COMBINATORIAL ;output enable
PIN 12..15  sum[3..0]       COMBINATORIAL ;intermediate sum
;pin 12=H indicates input>output pin 12=L indicates input<output
PIN 16      equal           COMBINATORIAL ;H indicates input=output
PIN 17      carry0          COMBINATORIAL ;intermediate terms to
PIN 18      carry1          COMBINATORIAL ;make the design fit
PIN 19      carry2          COMBINATORIAL ;within the PALCE16V8
PIN 10      GND
PIN 20      VCC

EQUATIONS

sum[0] = input[0]:+:threshold[0] ;half adder for LSB,
sum=input[0]:+:/threshold[0]:+:1 minimised
   carry0 = input[0]+/threshold[0]
```

```
           ;carry from input[0]+/threshold[0]+1 minimised
sum[1] = input[1]:+:/threshold[1]:+:carry0 ;sum for bit 1
carry1 = (input[1]*/threshold[1]) + (input[1]*carry0) +
                        (/threshold[1]*carry0)
sum[2] = input[2]:+:/threshold[2]:+:carry1
    ;sum for bit 2, sum[2] accommodates overflow
carry2 = (input[2]*/threshold[2]) + (input[2]*carry1) +
                        (/threshold[2]*carry1)
sum[3] = input[2]:+:/threshold[2]:+:carry2
    ;input[3]=input[2] and threshold[3]=threshold[2]
    ;sum[3] is sign bit of result
equal = /sum[0]*/sum[1]*/sum[2]
    ;zero addition result means input=threshold
```

SIMULATION

```
SETF   /oe
SETF /input[2] /input[1] /input[0] /threshold[2] /threshold[1] /threshold[0]
CHECK /sum[3]   /sum[2]   /sum[1]   /sum[0]    equal ;0=0
SETF /input[2] /input[1]  input[0] /threshold[2] /threshold[1] /threshold[0]
CHECK /sum[3]   /sum[2]   /sum[1]    sum[0]   /equal ;1>0
SETF /input[2]  input[1]  input[0] /threshold[2]  threshold[1] /threshold[0]
CHECK /sum[3]   /sum[2]   /sum[1]    sum[0]   /equal ;3>2
SETF /input[2] /input[1]  input[0] /threshold[2]  threshold[1]  /threshold[0]
CHECK  sum[3]   sum[2]    sum[1]    sum[0]   /equal ;1<2
SETF  input[2]  input[1]  input[0]  threshold[2] /threshold[1]  threshold[0]
CHECK /sum[3]   /sum[2]   sum[1]    /sum[0]  /equal ;-1>-3
SETF  input[2] /input[1]  input[0]  threshold[2]  threshold[1]  threshold[0]
CHECK  sum[3]   sum[2]    sum[1]    /sum[0]  /equal ;-3<-1
SETF  input[2]  input[1] /input[0]  threshold[2]  threshold[1] /threshold[0]
CHECK /sum[3]   /sum[2]   /sum[1]   /sum[0]   equal ;-2=-2
```

Simulation shows correct operation

```
                  ggggggg

THRESHOLD[2]      LLLLLHHH
THRESHOLD[1]      LLLHHLHH
THRESHOLD[0]      LLLLLHHL
INPUT[2]          LLLLLHHH
INPUT[1]          LLLHLHLH
INPUT[0]          LLHHHHHL
OE                LLLLLLLL
SUM[3]            LLLLHLHL
SUM[2]            LLLLHLHL
SUM[1]            LLLLHHHL
SUM[0]            LLHHHLLL
EQUAL             HHLLLLLH
CARRY0            HHHHHHHH
CARRY1            HHHHLHLH
CARRY2            HHHHLHLH
```

6.4 Concluding comments and further reading

This chapter has introduced the main number coding systems used in computer and binary arithmetic. Coding systems find much wider use in communications systems to improve efficiency and for reasons of data security. The coding system always needs to be deciphered, so that we can tell exactly what the code meant. A hardware implementation will be preferred owing to its speed, though the complexity of the coding technique might require a software solution, perhaps even one implemented using a parallel computer system to achieve a required speed. The examples included here have only concentrated on simple techniques for basic code conversion, but form the basis for extension to more complex systems.

The basis of computer arithmetic has been introduced. The most versatile system is floating-point arithmetic, and this can be implemented in software in computer systems to handle the arithmetic processes. A hardware implementation is often preferred for reasons of speed. Many modern computer systems include a *floating-point processor* that is specifically designed to implement arithmetic. Some central processing units include a floating-point processor; for example, the T800 transputer chip used in parallel processing systems extended an earlier transputer chip by including floating-point arithmetic. These processors are beyond the scope of this text, which has concentrated on basic designs only. These have highlighted the duality that exists between design approaches to combinational logic. This is not unexpected, since these approaches are just different ways to reach the same solution. Alternative design approaches can become attractive, owing to the nature of the design tackled. One of the difficulties in the PAL implementations herein is that we have used AND/OR structures that are unsuited to implementing arithmetic. There are other PALs with EXOR gates instead of the OR gates (which are sometimes called arithmetic PALs) that would better suit this type of design. There are also more sophisticated design techniques for arithmetic circuits, though these are beyond the scope of this text.

With regard to further reading on coding systems, particularly those used in communications systems, for an introduction try Stremler, F. G., *Introduction to Communication Systems* (Addison-Wesley, 3rd edn, 1990) or Proakis, J. G., *Digital Communications* (McGraw-Hill, 2nd edn, 1989) for a more advanced coverage. If you would like to find more details concerning arithmetic circuits and their implementation Wilkinson, B. and Makki, R., *Digital System Design* (Prentice-Hall, 2nd edn, 1992) provide a more detailed introduction, particularly in the context of processor design, whereas Mead, C. and Conway, L., *Introduction to VLSI Systems* (Addison-Wesley, 1980) concerns aspects of VLSI implementation. Patterson, D. A. and Hennessy, J. L., *Computer and Organisation and Design – the Hardware/Software Interface* (Morgan Kaufman, 1993) provide an overview of modern implementations whereas Swartzlander, E. (ed.), *Computer Arithmetic*, vol. II (IEEE Computer Society Press, 1990) is very detailed with good coverage but is actually a collection of scientific papers and as such is suited to more advanced study. Treatment of arithmetic PALs can be found in Bolton, M., *Digital Systems Design with Programmable Logic* (Addison-Wesley, 1990). Finally, Muroga, S., *VLSI Systems Design* (Wiley, 1982) offers an alternative study of binary to BCD conversion and includes some interesting algorithms.

6.5 Questions

1 Design a circuit to convert a 3-bit natural binary number to a 3-bit Gray code binary number

2 A circuit is required which divides two 2-bit binary numbers AB and CD to provide a 4-bit binary result WX·YZ (AB/CD = WX·YZ). The result is fixed point, with two bits for the integer part WX and the fractional part YZ. Design a circuit to implement this function and implement it using standard combinational circuitry. Implement it again in a PAL and show that the results are the same.

3 Design a circuit using basic binary subtraction that subtracts ten (expressed as a binary number) from a 4-bit binary number.

4 Design a circuit using one's complement arithmetic that subtracts a 4-bit binary number from ten expressed as a binary number (assume the 4-bit binary number $> 10_{10}$).

5 A circuit is required to convert single 4-bit numbers (between 0 and 9) to a coded output form as:

Input number	5-bit output
0	0 0 0 1 1
1	0 0 1 0 0
2	0 0 1 0 1
3	0 0 1 1 0
4	0 0 1 1 1
5	0 1 0 0 0
6	0 1 1 1 1
7	1 0 0 0 1
8	1 0 0 1 1
9	1 0 1 0 1

After examining this table, construct a software statement that expresses how the input is converted to the output. Implement the circuit using:

(a) a memory chip;
(b) an adder, a digital comparator and 2-1 line multiplexers; and
(c) using a PAL.

Compare and contrast your solutions.

6 A circuit is required to add or subtract two 2-bit binary numbers A and B where a mode signal M selects addition or subtraction. The signal M is LOW (M = '0') to select addition, to provide A + B, and HIGH (M = '1') to select subtraction, to provide A – B. The result is to be provided as a 3-bit binary number (assume A ≥ B).

Design a circuit to implement this function by using

(a) basic binary arithmetic;
(b) two's complement arithmetic; and
(c) confirm your results are the same for (a) and (b) for one bit of the 3-bit result.

7 Analog-to-digital and Digital-to-analog Conversion

*'Change is not made without inconvenience,
even from worse to better.'*

Richard Hooker

7.1 Analog-to-digital and digital-to-analog conversion basics

Digital circuits, and computers in particular, are often used to interpret signals to provide some form of control action. This can include **process control**, e.g. where the thickness of a metal strip is to be controlled. The computer or digital circuit is supplied with measurements of thickness from a *transducer* that provides an output voltage proportional to the strip thickness. This is used to control the pressure of the rollers on the metal strip to control its thickness:

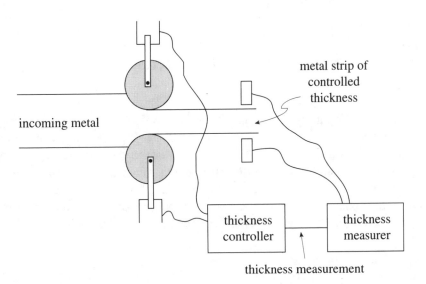

incoming metal

metal strip of
controlled
thickness

thickness
controller

thickness
measurer

thickness measurement

Signals are usually available only in *continuous* or *analog* form, since there are few transducers that provide a direct digital output. The analog voltage then needs to be converted to a digital number, so that it can be used by the digital circuitry or computer; this is achieved using an analog-to-digital (A/D) converter, which gives as output a digital number corresponding to the analog input. When the control signal has been worked out by the controller, it is in the form of a digital number and needs to be converted to give an analog signal; this is achieved using a digital-to-analog (D/A) converter.

A computer-based strip thickness controller is then of the following form:

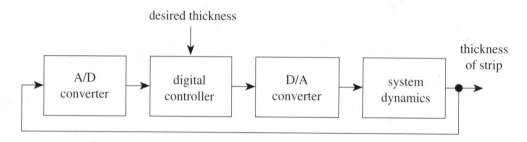

These production systems operate at a very impressive speed, but even this is still slow in comparison with the speeds associated with digital circuits. Faster applications include **speech synthesis** and **speech recognition** where computers produce and recognise speech. This requires a processing speed of the order of one hundred times faster than that for production control. An even faster technology is **computer vision** where computers are used to acquire and process television pictures.

To satisfy the wide range of differing requirements there is a choice of D/A and A/D converters with differing speed and performance characteristics. We shall look at these in general before we move on to examine the circuits used for A/D and D/A conversion.

7.2   A/D and D/A performance characteristics

As seen earlier, an n-bit binary integer has 2^n possible values. This is the same for an n-bit A/D or D/A converter. A 10-bit A/D converter splits the (input) analog voltage range into 2^{10} ranges, and assigns a digital code (or *discrete* value) to the nominal mid-range points. The number output by an A/D converter is a *sample* (instantaneous value) of the input analog signal. An 8-bit D/A converter can provide 2^8 discrete voltage values which correspond to the binary number fed as input to the converter. A transfer characteristic relates input to output; a 3-bit converter has only eight possible discrete ranges:

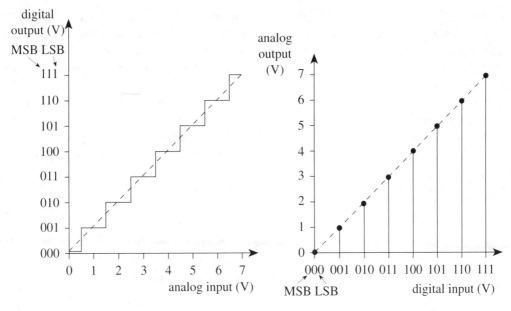

A/D converter characteristic   D/A converter characteristic

The *resolution* of an *n*-bit converter is *n* bits. The maximum analog voltage is the *full-scale voltage*; the smallest change that can be resolved is 2^{-n} of the full-scale span. For a D/A converter the maximum output voltage is specified by the chip manufacturer and choice often depends on application. The input voltage ranges for A/D converters are usually specified for a particular chip. Note that the origin is a usually a mid-range value, since the analog voltage might be negative as well as positive. The *absolute accuracy* is the difference between an output and its expected value and is often expressed in fractions of the least significant bit.

The *linearity* or *integral non-linearity* of a converter is the difference between a straight-line fit through the characteristic and the characteristic itself. The *differential linearity* is the maximum difference in step size from its expected value, and exists wherever there is integral non-linearity. Note that the components of A/D and D/A circuits can only be manufactured to a certain precision, which in turn dictates the precision of the converted signal.

The *conversion time* is of particular importance to D/A converters, and derives from the *switching time* (which derives from the delay before the converter responds), together with the *rise time* (which is how long the output takes to reach its final value), together with the *settling time* (which is how long before the output settles to a constant value). Note that A/D converters are often preceded with a *sample-and-hold circuit*, which ensures that the input voltage is held constant while conversion takes place. An *aperture* time defines how long before a signal, indicating conversion should start, takes effect. The conversion time of A/D converters is usually controlled by the number of clock cycles required before conversion is complete.

These are the basic performance characteristics of importance to a designer. Other characteristics that may be of interest, particularly in production, include *temperature co-*

efficients (whether the performance stays the same with variation in ambient temperature), *reference voltages* (many converters require reference or *offset voltages* and these can increase circuit cost), and the range of possible power supplies. Some D/A converters provide a current proportional to the input binary code, and this can easily be converted to a voltage using a resistor.

Finally, a topic of major importance in sampled data systems is the *sampling frequency*. This must be sufficiently high for the samples to represent the sampled waveform (so that we could reconstruct the sampled waveform from them). The minimum sampling frequency is given by *Nyquist's sampling theorem*, which states that we must sample at at least twice the highest frequency component of the original signal. If we do not, then the samples suffer from *aliasing* and do not represent the original signal accurately.

> *Digital signal processing (DSP) is a major topic using sampling theory. This is used in communications systems and many other areas. There are chips dedicated to DSP alone.*

7.3   Amplifiers and operational amplifiers

D/A and A/D converters use amplifier circuits that can be constructed from an *operational amplifier* (or op-amp, so called for its use in performing mathematical operations), configured as an *inverting feedback amplifier*. As such, this is more the domain of electronic circuits and more information can be found in the selection of textbooks at the end of this chapter. Described very briefly, the operational amplifier has a high *gain* (amplification factor) applied to the voltage which appears between the **non-inverting** input (denoted V^+) and the **inverting** input (V^-). The op-amp has power supplies, and if the amplified voltage at the output aims to exceed these, then it will saturate and the output voltage will remain at the power supply voltage that it has reached. An op-amp is symbolised by a triangle containing the two inputs:

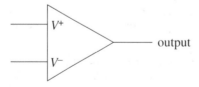

As such, the output voltage = $A(V^+ - V^-)$ where, say, the internal gain of the operational amplifier, $A = 10^5$. The op-amp can be used as an *analog comparator* providing an output which indicates whether the voltage applied at the non-inverting input V^+ is greater than the voltage applied at the inverting input V^-, or less than it. (A digital comparator, described earlier, is used to compare binary numbers.) The output will (by virtue of its large gain A) either saturate at the positive power supply for $V^+ > V^-$, or remain at the negative power supply for $V^+ < V^-$. This can then provide a logical output which is '1' when $V^+ > V^-$ and '0' when $V^+ < V^-$, respectively:

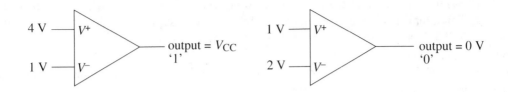

The op-amp can also be used to amplify the voltage applied at a single input. In an *inverting feedback amplifier* the non-inverting input is grounded and the inverting input is fed, via resistors, by the output and the input voltages:

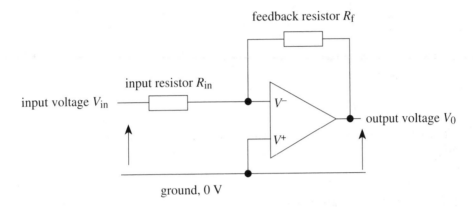

This configuration provides gain equal to the ratio of the two resistors. One resistor is attached between the input and V^-, while the other is a feedback resistor connecting the output to V^-. The non-inverting input is connected straight to ground, $0\,V$. Given that the op-amp has very high internal gain, the input V^- must be at nearly $0\,V$, since otherwise there would be a voltage difference between the inputs to the op-amp and the output would saturate. A voltage applied at the input is then dropped over the input resistor R_{in} and provides a current i_i. This current must be the inverse of the feedback current i_f through the feedback resistor R_f, or otherwise there would be net current into the inverting input which would then not be at $0\,V$. The operation then centres around the fact that the inverting input, is nearly $0\,V$, and this point is called *virtual earth* for this reason. The sum of currents into the virtual earth point and hence the non-inverting input V^- is then zero:

$$i_i + i_f = 0 \quad \text{and so} \quad i_f = -i_i$$

and since

$$V^- = 0\,V, \text{ then } i_i = V_{in}/R_{in} \text{ and } i_f = V_o/R_f$$

by substituting for i_f and i_i this gives

$$V_o = -R_f\, V_{in}/R_{in}$$

The output voltage V_o is then the input voltage, V_{in}, scaled by an amplification factor R_f/R_{in}. This amplification derives power from the power supplies feeding the operational amplifier. These are rarely shown on circuit diagrams, but actually provide the power in the output signal. By adding more inputs to the inverting feedback amplifier we obtain a *summing amplifier*, since the currents into the inverting input are summed and in total equal the fed back current because V^- is virtual earth:

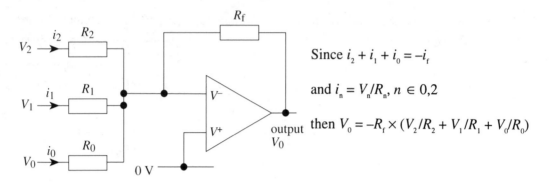

Since $i_2 + i_1 + i_0 = -i_f$

and $i_n = V_n/R_n$, $n \in 0,2$

then $V_0 = -R_f \times (V_2/R_2 + V_1/R_1 + V_0/R_0)$

These amplifier circuits are used in D/A converters, and analog comparators are found in A/D converters.

7.4   D/A conversion circuits

7.4.1   *Basis of a D/A converter*

A D/A converter aims to produce an output voltage proportional to an input binary number. The bits of the binary number can be weighted according to their significance. We can then derive a current proportional to the contribution of each bit by simply attaching the input bit to a resistor. We can then sum the currents through the resistors using a much smaller resistor that provides an output voltage, albeit rather a small one:

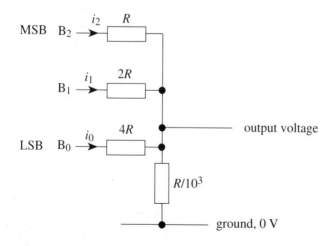

Given that a '1' = 5 V, then for a small output voltage if a bit is set then the whole of the 5 V will be dropped across its resistor. For each of the three resistors, the current if that bit is set is

$$i_2 = 5/R, \qquad i_1 = 5/2R, \qquad i_0 = 5/4R$$

The output voltage is then the sum of these currents through the output resistor:

$$V_o = (i_2 + i_1 + i_0) \times R/10^3$$

The current from the LSB is then divided by the highest factor and contributes least to the output. Conversely, the current corresponding to the MSB is greatest, since the voltage is divided the least and so the MSB contributes most to the output. Given $R = 10\text{k}\Omega$ then $i_2 = 0.5$ mA, $i_1 = 0.25$ mA, $i_0 = 0.125$ mA. For an input digital number $B_2 B_1 B_0 = 101$ then $V_o = 6.25$ mV and for an input 011, $V_o = 3.75$ mV.

These output voltages are very small, and this emphasises why this circuit only presents the basis of a D/A converter. Note that the weighting factor is the value of the resistors, and these are in binary order. The converter then forms the basis of a *binary-weighted converter*.

7.4.2 Binary-weighted D/A converter

The full binary-weighted converter includes a summing amplifier to ensure that the output voltage of the D/A converter is large and controllable:

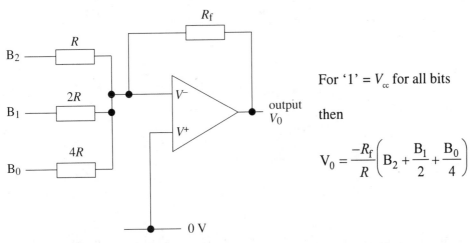

For '1' = V_{cc} for all bits

then

$$V_0 = \frac{-R_f}{R}\left(B_2 + \frac{B_1}{2} + \frac{B_0}{4}\right)$$

Binary-weighted D/A converter

A multiplying DAC is achieved by varying the reference voltage so that the output is then proportional to the product of the reference voltage and the binary input.

The major difficulty with this circuit is that as the number of bits increases, so does the size of the resistors. For an *n*-bit converter we need resistor values increasing in binary sequence from R, $2R$, $4R$, $8R$, $16R$, $32R$ to $2^{n-1}R$. For a 10-bit converter the LSB is weighted by a resistor 512 times larger than that weighting the MSB. In a chip this resistor is physically much larger and hence requires more space. It is also very difficult to manufacture resistors of such ratios to good precision. For these reasons a different circuit is usually preferred.

7.4.3   R/2R ladder D/A converter

The *R/2R ladder* is a circuit which only uses resistors of value R and $2R$. The binary inputs control switches which are connected either to V_{cc} when the bit is a '1', or to 0 V when the bit is '0'. The R/2R ladder then sums the currents derived from the switches and amplifies the result using an inverting feedback amplifier. Its main advantage is that the large ratios of resistor values associated with a binary-weighted converter are not needed:

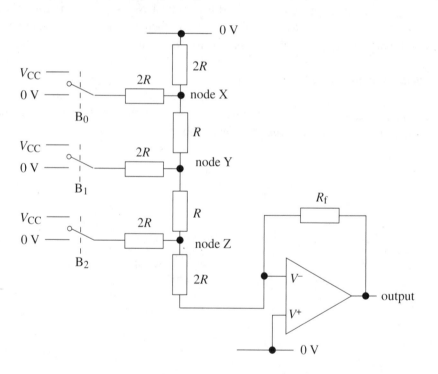

To analyse this circuit the main point is that the current injected by each input splits at each node. This is because the resistance seen at each node is always $2R$ upwards or $2R$ downwards (noting that the input to the inverting input V^- is virtual earth). Looking up or down from node Y with B_1 set and $B_2 = B_0 = $ '0' (and their inputs are then 0 V), we see a resistance R in series with a parallel combination of two $2R$ resistors. The resistance is the same both up or down and any current injected by B_1 will split along either branch. At node Z the current will see a resistance $2R$ down to the virtual earth point and $2R$ to the grounded input B_2 and will then split in half again.

If $B_0 = B_1 = $ '0' and $B_2 = $ '1', then current injected at point Z will see a resistance $2R$ down and upwards a resistance R in series with a parallel combination of R in series with two $2R$ resistors in parallel (using + to denote a series connection and || for parallel, this combination upwards from Z is $R + 2R \parallel \{R + 2R \parallel 2R\}$). The resistance seen upwards is then $2R$, so the current again splits at this node. The analysis for B_0 is the same (by reflection) for current injected into node X. Consequently, any current entering at each node splits equally between each path. The current injected by B_2 splits once only, current injected by B_1 splits in half twice, and the current injected by B_0 is split in half at each of the three nodes. By superposition, the current summed through the inverter is then the current injected by each bit $i(B_n)$ weighted according to the number of nodes at which it split

$$i_f = - (i(B_2) + i(B_1) + i(B_0)) = - (i/2 + i/4 + i/8)$$

Since the current splits along each path at a node, the equivalent circuit for B_2 is a $2R$ resistor in series with a resistor R. The current i_2 injected by B_2 is then $i_2 = B_2/3R$. This current is split in two and so its contribution to the output voltage is

$$V_o(B_2) = -\frac{R_f}{3R}(B_2 / 2)$$

The currents injected by B_1 and B_0 are the same as i_2, but split more times, so the total output is, by superposition, the voltage V_o due to the current injected by all three bits as

$$V_o = -\frac{R_f}{3R}\left(\frac{B_2}{2} + \frac{B_1}{4} + \frac{B_0}{8}\right)$$

The output then achieves a binary-weighted sum of the input bits. Note that the values of the resistors and the impedance of the switches will control the linearity that the converter can achieve. The switching speed will depend on the performance of the op-amp and on parasitic capacitance within the circuit.

7.5   A/D conversion circuits

7.5.1   *Single ramp A/D converter*

The simplest A/D converter is based on using a binary counter to count up until the analog equivalent of its count value just exceeds the input analog signal that is being converted. The arrangement uses a D/A converter to provide the analog signal for comparison with the input signal. A comparator provides a signal indicating whether the count should continue, since the count has not yet exceeded the input, or stop, when the output just exceeds the input:

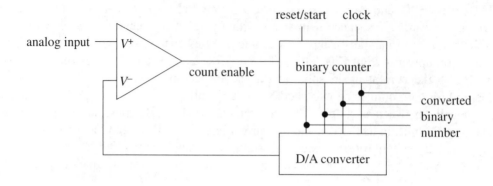

Conversion starts with the counter cleared. The output of the D/A converter is 0 V so the count enable signal will be '1', enabling counting. Counting will continue while the analog input is greater than the voltage at the non-inverting input, V^-. The counter will then increment at each clock cycle until the output of the D/A converter is slightly more positive than the input voltage. At this point the count enable signal will become '0', preventing counting, and the binary equivalent of the analog input is the output of the counter, which is then the output of the A/D converter. Since the output ramps to the input, the converter is called a *single-ramp converter*.

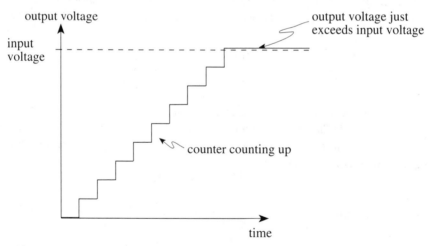

Single-ramp A/D converter

This method of A/D conversion process is rather slow. This is because we must allow of the order of 2^n clock cycles for an n-bit converter, for if the input is at its maximum then we have to count through the whole count sequence. For a 1 kHz clock a 10-bit converter requires of the order of 1024 ms to convert each input sample. Its advantage is that it is fairly simple in design.

> *Conversion usually starts by issuing a command and finishes with a signal indicating that conversion is complete. These are separate from counting and a single ramp converter would then require $2^n + 2$ clock cycles.*

7.5.2 Tracking A/D converter

The *tracking A/D converter* is a variant of the single-ramp A/D converter whose output is arranged to follow or track the analog input. This is achieved by replacing the (up) counter with an up/down counter and by using the output of the comparator as signal, indicating whether to count up or down. The counter should count up when the analog input is greater than the fed-back voltage derived from the counter output. The counter should count down when the fed-back voltage exceeds the analog input.

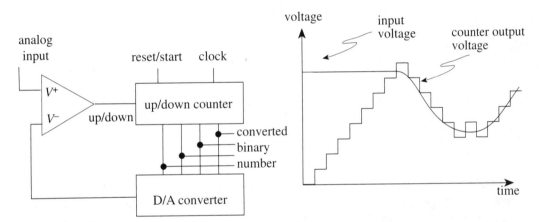

Tracking A/D converter

Note than the tracking A/D converter can only follow signals that change slower than the equivalent of 1 bit/clock cycle. If the analog input changes faster than this, then the tracking converter will suffer from *slope overload* when the output tries to reach the input voltage, but is an inaccurate representation of it.

> *The tracking A/D converter is the basis of a system known as delta modulation that is used in communication systems to reduce the amount of data transmitted.*

7.5.3   *Successive-approximation A/D converter*

7.5.3.1   *Successive-approximation converter*

The successive-approximation converter is a much faster A/D converter. It operates again by comparing the analog input with the output of a register fed via a D/A converter. The register is programmable, and we can set or reset each bit within it. At the end of conversion, the programmable register contains the converted value of the analog input. Each bit in the programmable register is set successively, starting with the MSB. The analog input is compared with sums of fractions of full scale, starting with the MSB. We start with the largest fraction, the MSB, which is half full scale and work through the bits successively until we reach the LSB, rejecting any bits that when set make the fed-back voltage greater than the analog input. The converter then approximates the input and the closeness of the output to the original signal increases as more bits are taken. For this reason the converter is called a *successive-approximation A/D converter*.

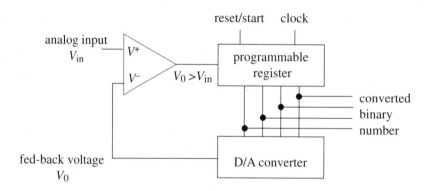

Consider that we have a 4-bit successive-approximation converter arranged to digitise inputs in the range 0 V to 16 V. Consider that the analog input is 11.2 V and all bits in the programmable register are reset. At the first clock cycle we set the MSB, B_3, of the converter, which provides an output V_o of 8 V, or half the 16 V full-scale span. The fed-back output voltage V_o is then less than the analog input, so the MSB is retained, $B_3 = $ '1'. The next bit, B_2, is set, and this corresponds to 4 V, one quarter of the full-scale range. The fed-back voltage now increases by 4 V to 12 V and is greater than the input, so this bit is not retained, $B_2 = $ '0'. The next bit B_1 is then set, so the fed-back voltage is 10 V, corresponding to the 8 V from B_3 plus the 2 V from B_1 corresponding to one eighth of the full scale. The fed-back voltage is less than the analog input, so B_1 is retained. The LSB, B_0, is then set so the fed-back voltage increases from 10 V to 11 V, corresponding to B_0 plus B_1 plus B_0. This is just less than the analog input, so B_0 is retained. All four bits have been examined, so conversion terminates with the final code $B_3B_2B_1B_0 = 1011$, corresponding to 11/16 of the full-scale span as required.

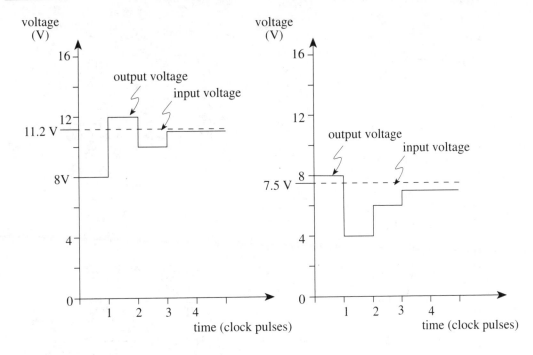

A second example shows conversion of an analog input just less than half of full scale. The MSB is not retained, since this takes the fed-back voltage too far to 8 V. The fed-back voltage then falls to 4 V, corresponding to $B_2 = \text{‘1’}$, and all other bits are reset. B_2 is retained and so are the next two bits, since neither takes the fed-back voltage above the analog input. Conversion terminates with the final code $B_3B_2B_1B_0 = 0111$, corresponding to an output of 7 V as required.

The main advantage of this approach is speed. An n-bit converter requires of the order of only n clock cycles for conversion, whereas the single-ramp converter required 2^n clock cycles. There are successive approximation converters available with conversion times up to a conversion period of 1 μs. Faster conversion requires a different technology. The problem remains – how do we design the programmable register? Many textbooks omit this topic, but it is actually quite an interesting problem in sequential design, and there are several approaches that we can use. We shall start with the main formal technique, ASM, and the move to semi-formal techniques and contrast the results achieved with commercial VLSI design.

7.5.3.2 ASM programmable register design

The starting point for ASM design is of course the ASM chart. We shall assume that we have two control signals, one to indicate that conversion should start, the other to indicate that the fed-back output is greater than the analog input. This control signal is the comparator output, and is ‘1’ if the output is greater than the input and ‘0’ otherwise. The ASM chart for a 2-bit programmable register is then

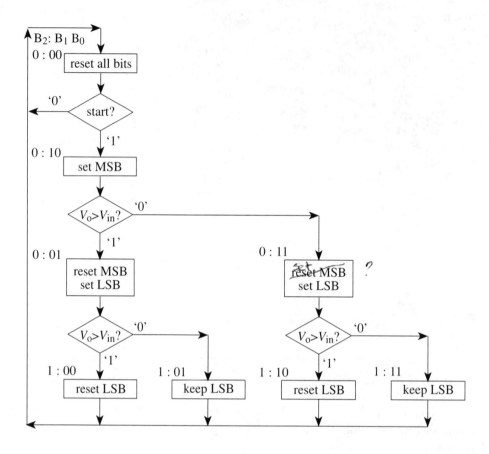

This gives the functionality that we require, starting with the MSB and finishing when the LSB has been tested. The ASM chart includes the bistable coding. With a 2-bit converter there are only four possible output codes and, for a number with bits in order MSB LSB, the codes are 00, 01, 10 and 11, which correspond to 0, ¼, ½ and ¾ of full scale, respectively. We can use these in the ASM chart as part of the state allocation. There are eight states in the ASM chart, however, so we need three bistables to implement it. The first bit, B_2, can be used to indicate whether conversion is complete ('0' = not complete) and the other two bits, B_1 and B_0, are the bits set in the converted number. The state allocations shown on the ASM chart are then for a coding B_2: B_1B_0 = conversion complete: MSB, LSB. The first bit indicates conversion is complete and the other two give the appropriate binary code for the converted number. The design is then extracted from the present state/next state chart:

Present state			Start	$V_o > V_{in}$	Next state		
Q_2	Q_1	Q_0	S	V	D_2	D_1	D_0
0	0	0	0	X	0	0	0
0	0	0	1	X	0	1	0
0	1	0	X	0	0	1	1
0	1	0	X	1	0	0	1
0	0	1	X	0	1	0	1
0	0	1	X	1	1	0	0
0	1	1	X	0	1	1	1
0	1	1	X	1	1	1	0
1	X	X	X	X	0	0	0

and the extraction is

$$D_2 = \overline{Q}_2 \cdot Q_0$$
$$D_1 = \overline{Q}_2 \cdot \overline{Q}_1 \cdot \overline{Q}_0 \cdot S + \overline{Q}_2 \cdot Q_1 \cdot \overline{V} + \overline{Q}_2 \cdot Q_1 \cdot Q_0$$
$$D_0 = \overline{Q}_2 \cdot Q_1 \cdot \overline{Q}_0 + \overline{Q}_2 \cdot Q_0 \cdot \overline{V}$$

which gives a circuit for a 2-bit programmable register:

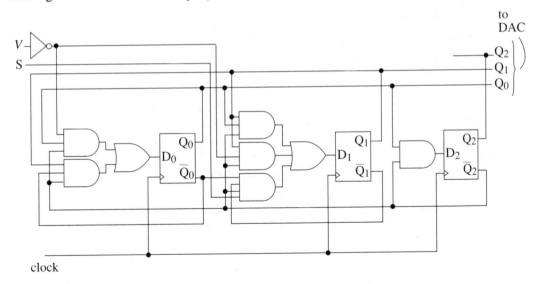

The minimisation actually needs a K-map for guaranteed minimisation, and so it is easier to use a PAL. The program is then

```
CHIP    programmable_register  PALCE16V8
;2-bit programmable register for successive approximation converter
;First define the pins
PIN  1     clk                              ;clock input
```

```
PIN   2      vo_gr_vi       COMBINATORIAL   ;1 if vout > vin
PIN   3      start          COMBINATORIAL   ;1 for start to convert
PIN   12     end_convert    REGISTERED      ;1 when conversion complete
PIN   13     msb            REGISTERED      ;msb programmable register
PIN   14     lsb            REGISTERED      ;lsb of programmable register
PIN   11     oe             COMBINATORIAL   ;output enable
PIN   10     GND
PIN   20     VCC


STATE ;define states and transitions
        MOORE_MACHINE

;define states according to successive approximation format
        none_set = /end_convert*/msb*/lsb
        test_msb = /end_convert*msb*/lsb
        smsb_test_lsb = /end_convert*msb*lsb
        rmsb_test_lsb = /end_convert*/msb*lsb
        smsb_slsb = end_convert*msb*lsb
        smsb_rlsb = end_convert*msb*/lsb
        rmsb_slsb = end_convert*/msb*lsb
        rmsb_rlsb = end_convert*/msb*/lsb

;transition equations state present to next state
        START_UP :=  POWER_UP -> none_set
        ;when power applied reset all registered pin outputs
        none_set := start -> test_msb ;start by testing msb
                   +-> none_set ;otherwise wait for start
        test_msb := /vo_gr_vi -> smsb_test_lsb ;then the lsb
                   + vo_gr_vi -> rmsb_test_lsb
        smsb_test_lsb := /vo_gr_vi -> smsb_slsb
                         + vo_gr_vi -> smsb_rlsb
        rmsb_test_lsb := /vo_gr_vi -> rmsb_slsb
                         + vo_gr_vi -> rmsb_rlsb
        smsb_slsb := VCC -> none_set      ;go to reset
        smsb_rlsb := VCC -> none_set      ;and wait
        rmsb_slsb := VCC -> none_set      ;to recommence
        rmsb_rlsb := VCC -> none_set      ;VCC = unconditional

SIMULATION
SETF /clk /oe
CLOCKF clk
CHECK /end_convert /msb /lsb
SETF  /start
CLOCKF  clk ;check reset state
CHECK /end_convert /msb /lsb
```

```
SETF start
CLOCKF clk   ;now set msb
CHECK /end_convert msb /lsb
SETF /start vo_gr_vi
CLOCKF clk   ;test msb and set lsb
CHECK /end_convert /msb lsb
SETF /vo_gr_vi
CLOCKF clk   ;save msb and test lsb
CHECK end_convert /msb lsb
CLOCKF clk   ;now return to reset
CHECK /end_convert /msb /lsb
SETF start
CLOCKF clk   ;try another sequence
CHECK /end_convert msb /lsb
SETF /vo_gr_vi
CLOCKF clk
CHECK /end_convert msb lsb
SETF vo_gr_vi
CLOCKF clk
CHECK end_convert msb /lsb
```

This uses a Moore machine to implement the state sequence. The coded values for the state are first defined and then the state transition diagram. The states are defined in terms of end_convert (for Q_2); MSB (for Q_1); and LSB (for Q_0). The Moore machine then expresses a set of state transitions and is labelled as such, since the start condition assumed on power up is the reset state, where all registered elements are reset. This is an alternative to the registered representation used earlier, but can become cumbersome when a large series of states are defined with specified outputs. It does express the state transitions more clearly and can be much reduced if the outputs are not labelled. The next state depends on the current state of the inputs, in this case the START signal and vo_gr_vi signal, which indicates that the fed-back output exceeds the analog input voltage. The symbol +-> indicates a local default for the reset state; if the start signal is not HIGH then the device remains in the reset state. The use of VCC implies an unconditional state transition. Since no outputs are defined for this state machine (the registered outputs are the outputs), it is essentially the same as the ASM specification.

This specification is compiled and achieves the same result as the ASM extraction (it should do – its definition is functionally identical) and simulates successfully.

```
LSB   :=   /VO_GR_VI * /END_CONVERT * LSB
      +   /END_CONVERT * MSB * /LSB
MSB   :=   /END_CONVERT * MSB * LSB
      +   /END_CONVERT * MSB * /VO_GR_VI
      +   START * /END_CONVERT * /MSB * /LSB
END_CONVERT   :=   /END_CONVERT * LSB
```

The waveform diagram is

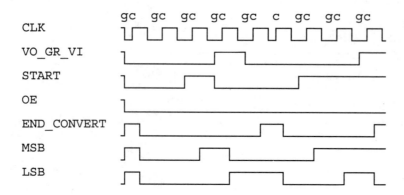

   The major difficulty with ASM design is its extension to large systems. A 2-bit programmable register is of little use in practical systems. An 8-bit register would be very
complex for ASM design, since the ASM chart would have 512 states, as the converter
would require an extra bit, used in part to indicate that conversion is complete. It is possible
using PALs, however, since the minimisation is done for us. To illustrate the problems with
ASM design we will specify a 4-bit converter for a PAL for outputs register[3] (MSB)
to register[0] (LSB), with a fifth bit register[4] to indicate that conversion is
complete. This becomes very complex, because there are now 32 states in the ASM chart.
The equations driving the register grow as well. The full coding is not included here, just the
pin and state specification and the result.

```
CHIP    4-bit_programmable_register   PALCE16V8

;2-bit programmable register for successive approximation converter

;First define the pins
PIN  1    clk                            ;clock input
PIN  2    vo_gr_vi      COMBINATORIAL    ;1 if vout > vin
PIN  3    start         COMBINATORIAL    ;1 for start to convert
PIN  12   end_convert   REGISTERED       ;1 when conversion complete
PIN  13   b3            REGISTERED       ;msb programmable register
PIN  14   b2            REGISTERED       ;lsb of programmable register
PIN  15   b1            REGISTERED       ;lsb of programmable register
PIN  16   b0            REGISTERED       ;lsb of programmable register
PIN  11   oe            COMBINATORIAL    ;output enable
PIN  10   GND
PIN  20   VCC

STATE ;define state transitions
```

```
        MOORE_MACHINE

;now define the states
        none_set = /end_convert*/b3*/b2*/b1*/b0
        test_b3 = /end_convert*b3*/b2*/b1*/b0
        sb3_test_b2 = /end_convert*b3*b2*/b1*/b0
        rb3_test_b2 = /end_convert*/b3*b2*/b1*/b0
        sb3sb2_test_b1 = /end_convert*b3*b2*b1*/b0
        sb3rb2_test_b1 = /end_convert*b3*/b2*b1*/b0
        rb3sb2_test_b1 = /end_convert*/b3*b2*b1*/b0
        rb3rb2_test_b1 = /end_convert*/b3*/b2*b1*/b0
        sb3sb2sb1_test_b0 = /end_convert*b3*b2*b1*b0
        sb3sb2rb1_test_b0 = /end_convert*b3*b2*/b1*b0
        sb3rb2sb1_test_b0 = /end_convert*b3*/b2*b1*b0
        sb3rb2rb1_test_b0 = /end_convert*b3*/b2*/b1*b0
        rb3sb2sb1_test_b0 = /end_convert*/b3*b2*b1*b0
        rb3sb2rb1_test_b0 = /end_convert*/b3*b2*/b1*b0
        rb3rb2sb1_test_b0 = /end_convert*/b3*/b2*b1*b0
        rb3rb2rb1_test_b0 = /end_convert*/b3*/b2*/b1*b0
        sb3sb2sb1sb0 = end_convert*b3*b2*b1*b0
        sb3sb2sb1rb0 = end_convert*b3*b2*b1*/b0
        sb3sb2rb1sb0 = end_convert*b3*b2*/b1*b0
        sb3sb2rb1rb0 = end_convert*b3*b2*/b1*/b0
        sb3rb2sb1sb0 = end_convert*b3*/b2*b1*b0
        sb3rb2sb1rb0 = end_convert*b3*/b2*b1*/b0
        sb3rb2rb1sb0 = end_convert*b3*/b2*/b1*b0
        sb3rb2rb1rb0 = end_convert*b3*/b2*/b1*/b0
        rb3sb2sb1sb0 = end_convert*/b3*b2*b1*b0
        rb3sb2sb1rb0 = end_convert*/b3*b2*b1*/b0
        rb3sb2rb1sb0 = end_convert*/b3*b2*/b1*b0
        rb3sb2rb1rb0 = end_convert*/b3*b2*/b1*/b0
        rb3rb2sb1sb0 = end_convert*/b3*/b2*b1*b0
        rb3rb2sb1rb0 = end_convert*/b3*/b2*b1*/b0
        rb3rb2rb1sb0 = end_convert*/b3*/b2*/b1*b0
        rb3rb2rb1rb0 = end_convert*/b3*/b2*/b1*/b0
```

with the result

```
    B0   :=  /VO_GR_VI * /END_CONVERT * B0
         +  /END_CONVERT * B1 * /B0
    B1   :=  /END_CONVERT * B1 * B0
         +  /VO_GR_VI * /END_CONVERT * B1
         +  /END_CONVERT * B2 * /B1 * /B0
    B2   :=  /END_CONVERT * B2 * B0
         +  /VO_GR_VI * /END_CONVERT * B2
```

```
        +   /END_CONVERT * B2 * B1
        +   /END_CONVERT * B3 * /B2 * /B1 * /B0
    B3  :=  /END_CONVERT * B3 * B0
        +   /END_CONVERT * B3 * /VO_GR_VI
        +   /END_CONVERT * B3 * B2
        +   /END_CONVERT * B3 * B1
        +   START * /END_CONVERT * /B3 * /B2 * /B1 * /B0
    END_CONVERT  :=  /END_CONVERT * B0
```

This shows that the product terms have grown with the complexity of the device. A hand-crafted ASM technique would clearly require a logic minimiser to reach a workable solution.

7.5.3.3 *Large programmable register design*

To handle device complexity designers often modify formal techniques to aid device design. This usually requires a lot of design experience. The major difficulty in ASM specification of large systems is a greater complexity, both in the specification and in the result. This can be handled by breaking the problem down into much smaller problems and finally assembling the subunits into a whole. It might be the case that ASM specification achieves a better-minimised result, but its complexity means not only that this is difficult to achieve but also that the result is difficult to maintain or debug.

The complexity of designing a large programmable register can be reduced by examining the functionality of the device and looking for an iterated solution. For each bit we first set it and then test it to see whether it took the output too far, or not. This is then repeated in succession until the LSB has been tested. The same cycle is used for each bit:

(a) set the bit;
(b) test to see if V_0 was greater than V_i when the bit was set;
(c) latch the bit for the remainder of the conversion process.

If we are provided with signals addressing each bit, for a 4-bit programmable register these signals are test_B_3 (the MSB), test_B_2, test_B_1, test_B_0 (the LSB), which are enabled ('1') when the appropriate bit is to be tested. These signals can be used to implement the testing cycle for each bit i ($i = 3, 2, 1, 0$):

(a) set B_i when test_B_i = '1';
(b) test B_i when test_B_{i-1} = '1' (the next least significant bit is enabled);
(c) latch B_i when neither B_i nor B_{i-1} is enabled.

This can be expressed in an ASM chart for each bit i as follows:

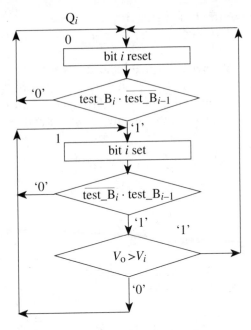

This can be summarised using a present state/next state chart as

Q_i	test_B_i	test_B_{i-1}	$V_o > V_i$?	D_i	
X	1	0	X	1	set B_i
X	0	1	0	1	setting B_i did not take us too far, keep it
X	0	1	1	0	setting B_i took us too far, reset it
0	0	0	X	0	keep reset bit
1	0	0	X	1	keep set bit

Each stage can then be implemented as a D-type bistable as follows:

$$D_i = \text{test\_}B_i \; + \; \text{test\_}B_{i-1} \cdot \overline{V_o > V_i} \; + \; \overline{\text{test\_}B_i + \text{test\_}B_{i-1}} \cdot Q_i$$

    set bit   now test it   and latch it in all other cases

which is implemented for each stage as follows:

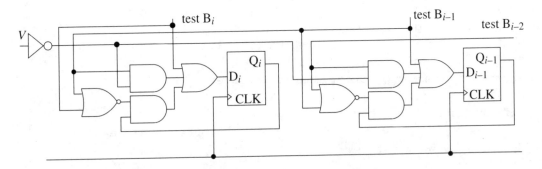

This could form the basis of a PAL description which for the first bit is

```
register[1] = test_bit[1] + test_bit[2].comp
                + \(test_bit[1] + test_register[2]).register[1]
```

We then need a circuit to provide a signal indicating which bit should be tested. This can be achieved using a shift register. If the shift register is provided with a pulse when conversion commences, then this pulse will pass through the shift register, enabling each bit in sequence. When this is included with the circuits setting and resetting each bit, then a 2-bit converter is given as

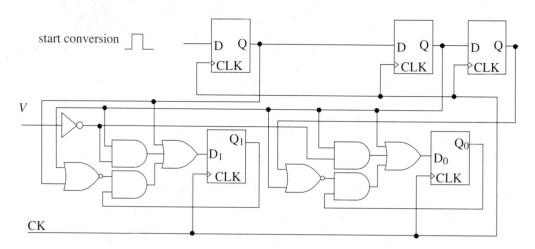

For a PAL the registered outputs of the shift register could be defined as shown earlier, and this can be included within a PAL with the earlier specification for testing and setting each bit in the programmable register. This would essentially be performing the logic extraction that PALASM should achieve. There are actually many ways of specifying the design, all of which lead to the same functionality, though with some differences in extraction. One specification is:

```
CHIP    programmable_register  PALCE16V8
;3-bit programmable register for successive approximation converter
;First define the pins

PIN  1        clk                          ;clock input
PIN  2        vo_gr_vi        COMBINATORIAL ;1 if vout > vin
PIN  3        start           COMBINATORIAL ;1 for start to convert
PIN  12..15   pregister[4..1] REGISTERED
;registered outputs of programmable register
PIN  16..18   shift[2..0]     REGISTERED
;shift register to enable each stage
PIN  11       oe                           ;output enable
PIN  10       GND
PIN  20       VCC
```

```
EQUATIONS
;register coding is [3:2:1] = [MSB:B2:B1]
;shift register is [2:1:0] where synchronised start signal enables
; pregister[3] , /shift[2] tests pregister[3],
; /shift[1] tests pregister[2], /shift[0] tests pregister[1]
; and pregister[4] signals the end of convert

shift[2] = start ;define shift register to enable each state
shift[1..0] = shift[2..1] ;latch data and shift other bits

;now look at programmable register bits in sequence
IF (/start)
     THEN BEGIN pregister[4] = 0       ;set end of conversion low
                pregister[3] = 1       ;set msb
                pregister[2..1] = 0   ;reset other bits
          END
     ELSE
     BEGIN    ;now test and reset bits appropriately
     pregister[3] = /shift[2]*/vo_gr_vi +
                         (shift[2])*pregister[3]
     pregister[2] = /shift[2] + /shift[1]*/vo_gr_vi +
                         (shift[2]*shift[1])*pregister[2]
     pregister[1] = /shift[1] + /shift[0]*/vo_gr_vi +
                         (shift[1]*shift[0])*pregister[1]
     pregister[4] = shift[2]*shift[1]
     END
pregister[4..1].CLKF = clk
shift[2..1].CLKF = clk


SIMULATION

SETF  /clk /oe
SETF  /start
CLOCKF clk
CHECK /pregister[4] pregister[3] /pregister[2] /pregister[1]
CHECK /shift[2]  shift[1] shift[0]     ;set MSB
SETF    start vo_gr_vi
CLOCKF clk
CHECK /pregister[4] /pregister[3] pregister[2] /pregister[1]
CHECK  shift[2]  /shift[1] shift[0]  ;reset MSB set B2
SETF    start /vo_gr_vi
CLOCKF clk
CHECK /pregister[4] /pregister[3] pregister[2] pregister[1]
CHECK  shift[2]  shift[1]  /shift[0] ;keep B2, set B1
SETF    start vo_gr_vi
```

```
CLOCKF clk
CHECK  pregister[4] /pregister[3] pregister[2] /pregister[1]
CHECK  shift[2]  shift[1]  shift[0]  ;reset B1, finish by setting p4
SETF  /start
CLOCKF clk
CHECK /pregister[4] pregister[3] /pregister[2] /pregister[1]
CHECK /shift[2]  shift[1]  shift[0]   ; start again
```

This compiles into

```
SHIFT[0]   :=  SHIFT[1]
SHIFT[1]   :=  SHIFT[2]
SHIFT[1].CLKF  =  CLK
SHIFT[2]   :=  START
SHIFT[2].CLKF  =  CLK
/PREGISTER[1]   :=  /START +  SHIFT[1] * /SHIFT[0] * VO_GR_VI
                +  SHIFT[1] * SHIFT[0] * /PREGISTER[1]
PREGISTER[1].CLKF  =  CLK
/PREGISTER[2]   :=  /START +  SHIFT[2] * /SHIFT[1] * VO_GR_VI
                +  SHIFT[2] * SHIFT[1] * /PREGISTER[2]
PREGISTER[2].CLKF  =  CLK
/PREGISTER[3]   :=  /SHIFT[2] * VO_GR_VI * START
                +  SHIFT[2] * /PREGISTER[3] * START
PREGISTER[3].CLKF  =  CLK
PREGISTER[4]   :=  SHIFT[2] * SHIFT[1] * START
PREGISTER[4].CLKF  =  CLK
```

This is very similar to the previous design. The means that to set and reset each bit according to the shift register is identical, but the latching circuitry differs from the ASM design. This is because the reset circuitry is now synchronous and depends on the current state of the system. This 4-bit register actually uses more output pins in the PAL, whereas an earlier 4-bit design only required four. The main advantage of the shift register/tester approach is that it provides a basis for VLSI designs by partitioning the design into manageable, and regular, elements. An alternative description, using vectors for this register is

```
CHIP    programmable_register  PALCE16V8
;3-bit programmable register for successive approximation converter
;First define the pins
PIN  1      clk                            ;clock input
PIN  2      vo_gr_vi        COMBINATORIAL ;1 if vout > vin
PIN  3      start           COMBINATORIAL ;1 for start to convert
PIN  12..15  pregister[4..1] REGISTERED
;registered outputs of programmable register
PIN  17..19  shift[2..0]     REGISTERED
;shift register to enable each stage
```

```
PIN   11      oe                              ;output enable
PIN   10      GND
PIN   20      VCC


EQUATIONS
;register coding is [3:2:1] = [MSB:B2:B1]
;shift register is [2:1:0] where synchronised start signal
;enables pregister[3], /shift[2] tests pregister[3],/shift[1]
;tests pregister[2], /shift[01] tests pregister[1] and pregister[4]
;signals the end of convert

shift[2] = start ;define shift register to enable each state
shift[1..0] = shift[2..1] ;latch data and shift other bits


;now look at programmable register bits in sequence
IF (/start*pregister[4]) ;when starting, set MSB, B3
     THEN BEGIN pregister[4] = 0         ;end_of convert =1
                pregister[3] = 1         ;set bit 3
                pregister[2..1] = 0      ;all other bits are reset
          END
     ELSE
     BEGIN
       CASE (shift[2..0],pregister[4])
       BEGIN
       #b0110: BEGIN pregister[4] = 0
                pregister[3] = /vo_gr_vi     ;test bit 3
                pregister[2] = 1             ;set bit 2
                pregister[1] = pregister[1]
            END                              ;latch other bits
       #b1010: BEGIN pregister[4] = 0
                pregister[3] = pregister[3] ;latch bit 3
                pregister[2] = /vo_gr_vi    ;test bit 2
                pregister[1] = 1            ;set bit 1
            END
       #b1100: BEGIN pregister[4] = 1
                pregister[3..2] = pregister[3..2]
                pregister[1] = /vo_gr_vi    ;test bit 1
            END
       #b0010,
       #b0100, ;start illegal shift register states
       #b1000, ;with MSB set
       #b0000: BEGIN pregister[4] = 0
                pregister[3] = 1
                pregister[2..1] = 0
            END
```

```
        OTHERWISE: BEGIN pregister[4] = 1
                         pregister[3..1] = pregister[3..1] END
                    ;latch all outputs until next start
            END
        END
pregister[4..1].CLKF = clk
shift[2..0].CLKF = clk
```

which in simulation achieves

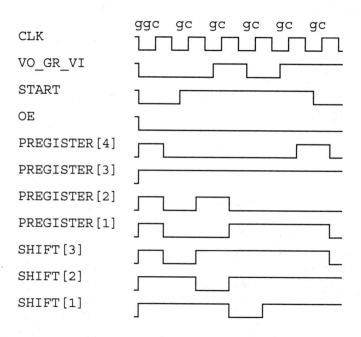

7.5.3.4 Comparison of techniques

Few data books reveal exactly what is inside the successive approximation register. Each stage of the NEC DM2504 programmable register uses SR latches; one registers the data, the other controls it. The control latches are negative-edge-triggered (via $\phi2$) and the data latches are positive-edge-triggered (via $\phi1$). Each stage is SET before it is tested, RESET when tested, and latched thereafter. The arrangement for the stages is

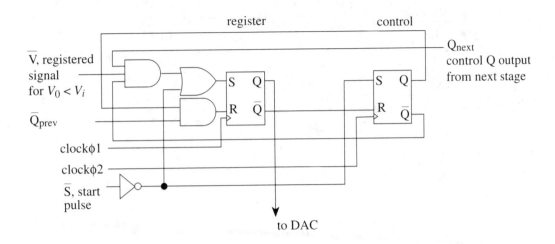

Each bit is tested with all remaining bits set, unlike the previous designs where a bit was '0' until tested. This is an alternative design approach which could have been used earlier. Conversion starts with the MSB LOW and all remaining bits HIGH. This is achieved by an active LOW pulse on the S input, which then sets the register bit HIGH (via the inverter), and the control output Q is also HIGH (via the inverter on its input data line). When each stage changes, the Q_{prev} input to the next stage goes HIGH, resetting the next register bit. This in turn resets the control Q output which, via a HIGH control Q output, on the next rising clock edge ($\phi 1$) latches the data, indicating whether the converted output voltage was less than the analog input voltage (or not), via the V input, into the register. When the control output Q (Q_{prev} for the next stage) changes HIGH, it resets the next register bit LOW ready for testing. Finally, the Q_{next} input, the control Q output from the next stage, ensures that the register latches thereafter. This can be compared with the successive approximation register earlier, which incorporated a shift register to test each bit. In the NEC DM2504 the output of the control latch of the previous stage initiates testing a register bit and the output from the next stage latches it, thereby including the functionality of the shift register within the programmable register. The device retains the property of successive approximation, but implements it in a different way. This could have been the design approach earlier, but an approach where a bit is reset until tested perhaps conforms better with the way we think about these systems.

A more compact version is much closer to a VLSI design. Here a shift register is used to enable, or set, each bit (as used in the previous section). The design is implemented using D-type bistables that register the comparator output in turn.

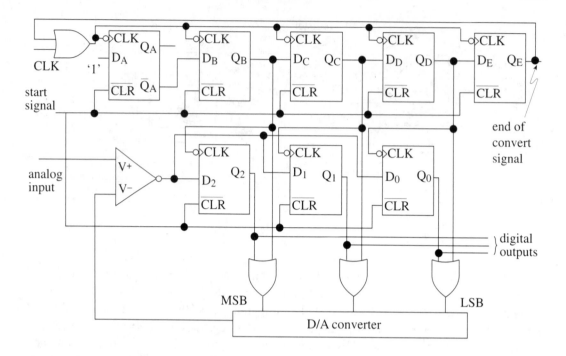

Conversion starts with a reset signal clearing the bistables. At the first falling edge of the clock, Q_A latches the '1' at its data input. Q_B latches a '1' at the first clock edge but is reset at the second. A pulse of one clock period then passes through the shift register. The output from the shift register is used to represent each bit being set in turn. The falling edge of the shift register outputs latch the comparator output into the register for each bit. After each bit has been tested and set it is then latched, since there is no further clock signal to activate change. The latched output bit is then fed to the D/A converter while the next bit is tested. Each bit is tested in succession as the convert pulse progresses through the shift register. When conversion is complete, the last bit of the shift register is SET, preventing a clock signal from triggering the shift register; conversion thus stops with the converted value held in the register.

Both of these successive approximation registers are **asynchronous**, since in neither system are all bistables synchronised to the same clock signal. While the circuits are much more compact than their synchronous counterparts, they are much more difficult to design. Though less effort is now directed towards the design of successive approximation A/D converters owing to the development of *sigma delta converters*, they still find applications in systems where performance up to a sampling frequency of 1 MHz is acceptable. A typical application is in the acquisition of human speech for purposes of recognition.

7.5.4 Flash A/D converter

The *flash* A/D converter is currently the fastest commercially available A/D converter. (It is not called flash because of its sophistication!) A flash converter achieves high speed by using a network of comparators, each fed with a reference level. If the input analog voltage is greater than its reference level, then the comparator is set. The highest comparator to be set is encoded to provide the digital output.

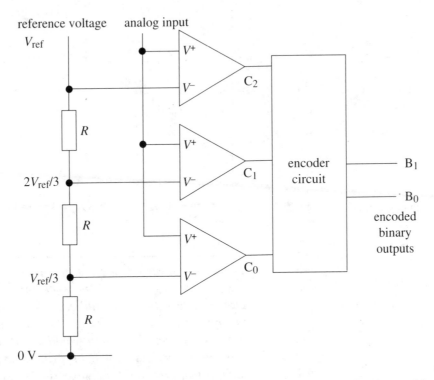

The flash converter actually works much like a thermometer. The reference levels are obtained by dividing up a reference voltage using a potentiometer chain. As the analog input voltage increases, more comparators are set and the highest level (cf. the temperature) is set by the analog input. Though an n-bit converter has 2^n levels, an n-bit flash converter contains only $2^n - 1$ comparators. This is because we do not need a comparator to tell us that the analog input is at its lowest value, since in this position no comparator is set (and that provides the information we need). The outputs of the comparators are encoded to provide the n-bit binary output. The network of comparators provides the high speed of operation. Response time is limited mainly by the speed of the comparators, together with the propagation delay of the encoder logic. The reference voltage is three quarters that of the full-scale input voltage, V_{fs}. Thus if the analog input is between $0\,V$ and $V_{ref}/3$ this is equivalent to setting the comparators when the analog input is in the range $0\,V$ to $V_{fs}/4$:

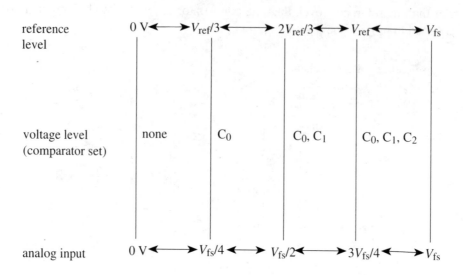

The design saves on one resistor as well as one comparator, by using a reference input that is three quarters that of the full-scale input range. If the reference voltage was set to equal the full range scale then the design would incur an extra resistor. The circuit would be a little simpler to understand, however (though a single resistor is rather inconsequential when compared with $2^n - 1$ comparators).

The coder design requires to provide an *n*-bit binary output, indicating the largest reference level of the comparators set by the analog input. For a 2-bit converter the encoder operates according to

Voltage level	Comparators set			Encoded binary output	
	C_2	C_1	C_0	B_1	B_0
0	0	0	0	0	0
1	0	0	1	0	1
2	0	1	1	1	0
3	1	1	1	1	1

One way of implementing B_1 and B_0 is to add the number of comparators that are set. This is because the highest comparator set indicates the level of the analog input. In the 2-bit converter, addition of C_2, C_1 and C_0 gives B_0 as the sum bit and B_1 as the carry. A better-minimised solution derived by a K-map is

$$B_1 = C_1 \qquad B_0 = C_0 \cdot \overline{C_2} \cdot C_1$$

This is actually an application of *priority encoding*, which involves devices that encode the position of the most significant bit set. An 8-bit priority encoder is described by its truth table as:

Position to be encoded							Priority encoder output		
B_6	B_5	B_4	B_3	B_2	B_1	B_0	D_2	D_1	D_0
0	0	0	0	0	0	0	0	0	0
0	0	0	0	0	0	1	0	0	1
0	0	0	0	0	1	X	0	1	0
0	0	0	0	1	X	X	0	1	1
0	0	0	1	X	X	X	1	0	0
0	0	1	X	X	X	X	1	0	1
0	1	X	X	X	X	X	1	1	0
1	X	X	X	X	X	X	1	1	1

For an n-bit converter, the single ramp converter required of the order of 2^n clock cycles, the successive approximation required n cycles and the flash converter requires the order of a single clock cycle. Flash A/D converters can achieve operating speeds of greater than 20 MHz. Their price fell sharply during the 1980s when many production difficulties were solved. For an 8-bit converter 255 comparators need to be integrated onto a single chip (and all offer a similar performance). The designer then has to compensate for temperature effects. Flash converters clearly predominate throughout high-speed designs. A typical application is in acquiring video television signals for computer vision systems.

7.5.5 Dual-slope A/D converter

The *dual-slope A/D converter* is a very accurate but slow A/D converter. Its basic operation is to use two integration stages, and it measures the time elapsed using a digital counter. The digital counter eventually contains a binary number corresponding to the analog input voltage.

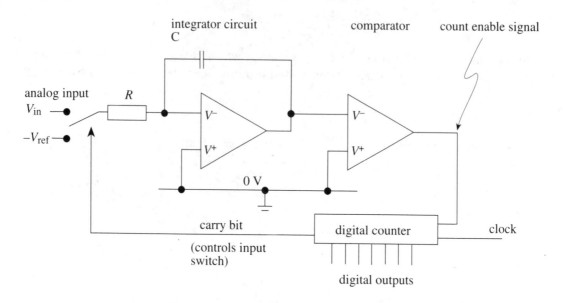

Conversion starts with the switch connected to the analog input, V_{in}. The output of the integrator is a ramp falling from 0 V with slope controlled by V_{in} (if we integrate a constant level, we get a ramp with time). This means that the output of the comparator is '1', since the inverting input is less than the non-inverting input. The output of the integrator will reach a terminal voltage V_t when the n-bit counter has counted through its full 2^n clock cycles. When this occurs the carry bit from the counter is set, which causes the switch to change from the analog input to the negative reference voltage. Since its input is now negative, the output of the integrator begins to rise from V_t. The counter starts to count through its cycle again. When the integrator output reaches 0 V the comparator output will become '0', thus preventing counting.

Since the point to which the integrator output fell is controlled by the analog input, V_{in}, and since the rate at which it rises is constant (controlled by V_{ref}), the time for the integrator output to reach 0 V depends on the magnitude of the analog input. Since counting stops when the integrator output reaches 0 V, the value held in the counter is the binary representation of the analog input voltage. The integrator output voltage will then fall faster for a larger input voltage than it does for a small input voltage. The terminal voltage, V_t, is then lower for large V_{in} than it is for a smaller V_{in}. After this point is reached, the integrator output then rises to 0 V at a rate irrespective of V_{in}, controlled by V_{ref}. In this manner, the number of clock cycles taken for the integrator output to reach 0 V depends on the magnitude of the analog input V_{in}. This can be summarised diagrammatically as follows:

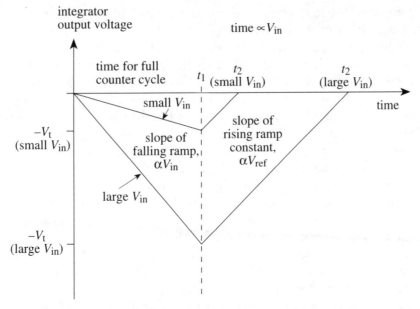

Mathematically, the integrator output returns to zero, having fallen according to the analog input from time 0 to time t_1, and then risen from t_1 from t_2 according to slope V_{ref}:

$$-\frac{1}{RC} \int_0^{t_1} V_{in} \, dt + \frac{1}{RC} \int_{t_1}^{t_2} V_{ref} \, dt = 0 \qquad \text{thus} \qquad V_{in} \times t_1 = V_{ref} \times (t_2 - t_1)$$

The time between the start and t_1 is 2^n clock cycles for an n-bit counter, so $t_1 = 2^n$. The time from t_1 to t_2 is m clock cycles; then

$$V_{in} \times 2^n = V_{ref} \times m \qquad \text{and thus} \qquad V_{in} \propto m$$

The number, m, of clock cycles counted by the converter in the second half of the cycle is therefore proportional to the analog input voltage. This is the binary number contained in the counter during counting, and hence conversion stops. The comparator signal is '0' at this point, and this can be used to indicate that conversion is complete.

The main advantage of the dual-slope converter is that any non-linearity in the first ramp is compensated for by the same non-linearity in the second ramp (since we use the same integrator). This can improve precision, but it also reduces speed. The dual-slope converter, though twice as slow as a single-ramp or single-slope converter, finds application in digital voltmeters where the poor speed can be tolerated but the precision is necessary.

7.6 Concluding comments and further reading

This chapter has reviewed various techniques which are used to convert analogue data to digital data and vice versa. The variety of circuits exists to satisfy many important performance characteristics. Though resolution and accuracy are naturally of great importance, since they govern the fidelity of the converted representation of the original signal, speed is of paramount importance too, since it governs whether a technique can be used at all.

As well as supporting study in the application of digital electronics, A/D converters also offer a rich source for study in logic design. This has involved the design of the successive-approximation converter, which shows ASM chart design in a practical application, and how ASM charts can be reduced by functional interpretation to lead to a more manageable design. Comparison with established commercial circuits shows how they operate by the same principles (as they must), but can use asynchronous design to achieve a more compact solution. This material is not included as a panacea for design, nor as an all-embracing study of data conversion, since many practical aspects are beyond the scope of an introductory text. It does, however, highlight some more practical aspects of digital systems design and their interconnection to process data from the 'real' world.

For an introduction to signal processing theory, Meade, M. L. and Dillon, R., *Signals and Systems* (Van Nostrand-Reinhold, 1986) provides an interesting introduction; for more detail and for more advanced treatment, Rabiner, R. L. and Gold, B., *Theory and Application of Digital Signal Processing* (Prentice-Hall, 1975) has proved a classic text, though there are many more recent ones. Signal processing theory is actually used in the design of sigma delta A/D converters and these are now the most exciting areas of development in data conversion; Candy, J. C. and Temes, G. C., *Oversampling Delta Sigma Converters* (IEEE Press, 1992) provides a very detailed picture. Seitzer, D., Pretzl, G. and Hamdy N. A., *Electronic Analog to Digital Converters* (Wiley, 1983) is one of few books dedicated to data conversion alone and reviews many different techniques (but even then it does not manage to discuss what is inside a successive-approximation converter), whereas a number of analog IC design books include it as a section; try for example Grebene, A. B., *Analog Integrated Circuit Design* (Van Nostrand-Reinhold, 1972). Haznedar, H., *Digital Microelectronics* (Benjamin, 1991) covers more technological implications of converter design. Manufacturers' handbooks remain the best source of reference on A/D and D/A converters: it is worth consulting in particular Analog Devices's *Data Converter Reference Manual* (vols I and II), Plessey's *Data Converters and Datacoms IC Handbook*, Texas Instruments's, *Linear Circuits Databook*, vol. 2, and those of specialist suppliers such as Maxim.

7.7 Questions

1 An 8-bit A/D converter is required to convert an analog input to the digital output within $1.0\,\mu s$. What technique would you choose to base this converter on?

2 An A/D converter is required to convert an analog input in the range 0 to 5 V to an accuracy of at least $10\,mV$. What resolution is required for the converter?

3 Implement part of a single ramp 8-bit converter using a PAL. What parts of the converter cannot be accommodated within the PAL?

4 Design a single-ramp converter that counts down to the digital output, rather than up.

5 In an R/2R ladder D/A-based converter (Section 7.4.3) the input resistor attached to B_2 (the MSB) is $0.9R\,\Omega$ rather than $R\,\Omega$. Explain what effect this might have, considering in particular the linearity and differential linearity of the converter.

6 Design a 2-bit successive approximation converter that starts with all bits set and tests each bit by resetting it. Conversion should start after a start signal is HIGH. What differences exist in function between this converter and one which operates by setting each bit to test it?

7 Design a flash A/D converter which converts an analog input voltage within the range $0\,V$ to $+5\,V$ with an accuracy of better than $35\,mV$.

8   Interface and Hybrid Circuits

> *'Dust as we are, the immortal spirit grows*
> *'Like harmony in music; there is a dark*
> *'Inscrutable workmanship that reconciles*
> *'Discordant elements, makes them cling together*
> *'In one society'*
>
> William Wordsworth, *Prelude*

8.1   Schmitt trigger

This chapter aims to introduce various circuits associated with digital design that have not found presentation earlier. Many of the circuits handle practical consequences of digital design, such as providing outputs with high current capability, or by handling noise that exceeds a noise margin. As such many of these circuits lie at the interface between pure digital and analog circuits. As such, the introduction here is phrased in digital terms, though you may want to study a conventional circuits text for the operation of some of these devices.

The first circuit we shall look at is the *Schmitt trigger*. We shall consider the *Schmitt trigger inverter* circuit, which has a **hysteresis property**. This property means that if the input voltage has increased beyond an upper switching threshold (causing a LOW output) then the input voltage needs to fall below the lower switching threshold before the output can switch HIGH. This is indicated on the characteristic by the arrow indicating a change from HIGH to LOW (depicted as right to left on the characteristic). Conversely, if the output is HIGH then the input needs to exceed the upper switching threshold before the output changes to LOW, as depicted on the characteristic as the arrow indicating switching HIGH to LOW (left to right).

There is a Schmitt trigger which uses a feedback circuit around an comparator to give the hysteresis effect.

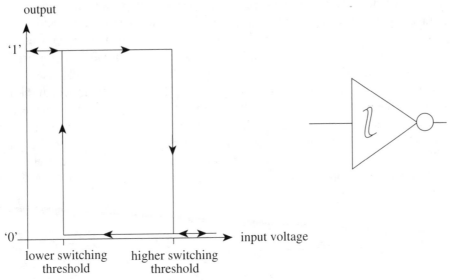

Schmitt trigger inverter characteristic   Schmitt trigger inverter symbol

The Schmitt trigger has many uses, especially as a circuit to receive noise-corrupted signals and to reduce the effect of noise on them (see Section 8.4). Another use of the Schmitt trigger is to **sharpen** slowly changing signals. These can cause problems in digital circuits, but if they are intercepted by a Schmitt-trigger-based circuit, the point at which the slowly varying signal passes the switching threshold will become a fast-moving edge between two well-defined logic levels. More gates than just an inverter are available with Schmitt trigger inputs.

One use of a Schmitt trigger is in a *reset* circuit which gives a signal, used to restart a sequential system. This is often achieved using a push-button. When the reset push-button is pressed the circuit restarts from the beginning of a sequence, but the bouncing action of the push-button can cause problems. The reset circuit can be more reliably achieved using a Schmitt-trigger-based inverter.

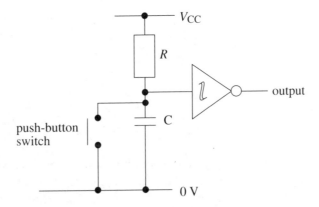

In this circuit, in a static case, before the reset push-button is pressed the capacitor will be charged, and the input voltage to the inverter will be HIGH, causing a LOW output. When the reset push-button is pressed, the capacitor discharges through it to ground, giving a LOW input to the inverter and hence a HIGH output. When the push-button is released, the capacitor charges up through the resistor and the input voltage will eventually rise to above the upper switching threshold and the output will switch LOW. The rate at which the input voltage rises (after the push-button is released) depends on the time constant of the resistor–capacitor network; this is the product of the resistance and the capacitance and is called the *time constant*. The length of the pulse can be controlled by appropriate choice of the resistor and capacitor values. This circuit does not provide a pulse of precisely controlled time length, owing to the variation in resistance and capacitance with time (capacitor leakage reduces capacitance with time).

8.2   Power supplies and decoupling

Constancy of logic levels output by a circuit clearly depends on the constancy of the power supplies which feed it. *Decoupling capacitors* can be introduced to reduce variation of the power supplies; these are capacitors connected between the positive and negative power supplies which provide a local reservoir of charge that is drawn as current, and they will top up the power supply if it tries to fall. These will reduce the effects of a *power supply glitch* caused by the transitional state of a TTL output circuit when, say, the pull-up transistor is switching ON and the transistor connecting the output LOW is switching OFF. This introduces an effective short-circuit, since the power supply is connected through two ON transistors to ground. This will consume excessive current, so a local decoupling capacitor will help to regulate the power supply and keep it constant. The power supply voltage will actually fall because of the resistance in the power distribution tracks and the output resistance of the power supply.

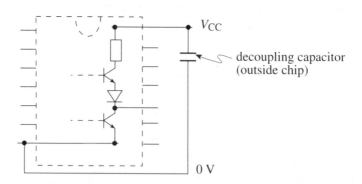

A basic power supply will actually include capacitors as part of the *smoothing* process converting an a.c. voltage supply into the steady d.c. voltage required as supply for digital circuits. By including decoupling capacitors, we are then distributing the power supply smoothing within the digital circuit. Decoupling capacitors are then placed as close as physically possible to a chip. There are actually several varieties of capacitors, each with advantages associated with a manufacturing technology. Small 100 nF **ceramic** and larger 10 μF **tantalum bead** capacitors are often used in application. Note that some capacitors are **polarised** and have a positive and a negative connection and hence need to be connected the right way round. **Voltage regulators** can also be used to improve d.c. power supplies, and there are **voltage converters** if you need to redeploy a particular supply voltage. There are circuits called *power supply supervisors*, which can provide a reset signal when the power supply voltage falls by an amount specified for correct circuit operation.

8.3   Output stages

8.3.1   Gate output circuits

There are several possible output stages for digital circuits. We have already considered the totem pole circuit in a TTL NAND gate, which employs a transistor to provide an active pull-up circuit, as well as the transistor to pull the output LOW. The advantage of this circuit is that it is a good compromise design. It has two main disadvantages: the power supply glitch (as discussed in Section 8.2), together with problems when gate outputs are connected together. If two totem pole outputs are connected together, everything is fine so long as all outputs are in the same state. If one output is HIGH and the other is LOW, then the point where the outputs are connected together will be somewhere between HIGH and LOW (there is also a large current with potential failure of the transistors), and most probably not in a valid logic state.

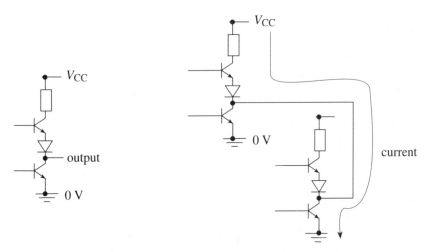

Totem pole output stage   Intermediate state resulting from totem pole output

The reason why we might want to connect outputs together is, for instance, when circuits drive a *data bus*, for example the *n* parallel data bits for an *n*-bit microprocessor. This data bus will accept signals from memory chips and from the outside world for processing. When the state of the bus is indeterminate, owing to competing output values, this is known as **bus**

contention. We then need some way to force a circuit to relieve the effect of its output. One way to accomplish this is to omit the active pull-up transistor in the TTL circuit. The collector is then disconnected, and for this reason these circuits are known as *open-collector* circuits. When the gates are deployed in circuits, the point where the outputs are connected together, i.e. the bus signal, is connected to the positive power supply using a single pull-up resistor. In this manner, if the output of any gate is LOW then the bus signal will be LOW. If the output of a gate is not LOW, and its output transistor is switched OFF, then it will have no effect on the bus signal. If the output transistors of all transistors connected to the bus signal are OFF, then the bus signal will be pulled HIGH by the pull-up resistor. In this manner only LOW gate outputs affect the bus signal, and so there is no bus contention.

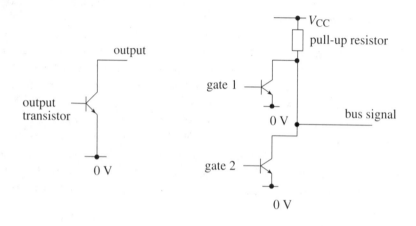

Open collector output   Open collector driving a bus signal

The connection is called *wired OR*, since if gate 1 OR gate 2 is LOW then the output is LOW. This actually depends on the convention we are using, since it can be called *wired AND*, because if gate 1 AND gate 2 are HIGH then the output is HIGH. Open collector logic is less popular now, since the choice in value of the pull-up resistor affects the speed that the circuit can achieve and its value needs to be chosen to ensure that current limits on the outputs are not exceeded. Open-collector logic is little used in PAL technology, although the PAL16V8HD can be programmed to provide an open-collector output by blowing a specified fuse.

We can exceed the maximum current by virtue of the wired OR configuration. If all open-collector outputs are OFF, then current will feed through the pull-up resistor into the following logic inputs. If some of the outputs are ON then the collector current is determined by the ON output resistance and the resistor value.

Tristate logic has largely superseded open-collector logic; its function is much easier to visualise and it is much easier to deploy. By its name, tristate logic has three states: ON; OFF; or *high impedance*. The extra state is the high-impedance state, usually denoted by Z in truth tables and in PAL programming systems. In this state the gate presents a large resistance to the bus and hence effectively disappears from the circuit since it draws no current. The high-impedance state is then equivalent to a **'not there'** state. There is usually a control signal enabling the circuit output, in which case it is ON or OFF, or disabling it when the output is high-impedance. It can be achieved in a CMOS inverter by using an extra pair of transistors to switch the output OFF (the device can be considered to be a double inverter). The ENABLE signal is inverted between the upper PMOS transistor and the lower NMOS transistor. Using both types of transistor is consistent with CMOS objectives and provides symmetry in performance. When the ENABLE signal is HIGH the circuit acts as an inverter since the extra transistors are ON. When the ENABLE signal is LOW, this presents a high-impedance output state.

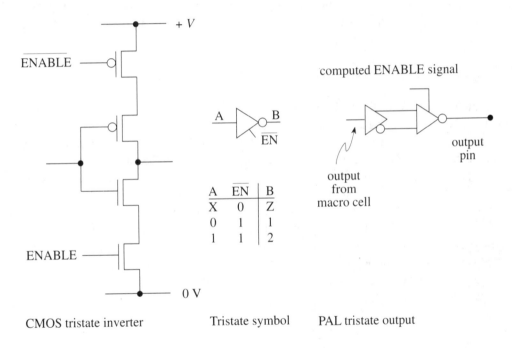

A	$\overline{\text{EN}}$	B
X	0	Z
0	1	1
1	1	2

CMOS tristate inverter   Tristate symbol   PAL tristate output

This enable signal can be used within PALs to enable circuit outputs.

8.3.2 *Enabling PAL outputs*

If the outputs are permanently enabled, then the output signals are allowed to appear; so, in the development stage, we can test the circuit using test vectors. In practice, designers often prefer to enable output signals only when they are required. The state coding internal to a state machine is often not required outside the PAL. The output stage is often a tristate buffer and when disabled it is in its high-impedance state, often allowing a pin to be used as an input connection. We can switch an output ON by specifying its enable signal to be HIGH:

```
output.TRST = VCC   ;permanently enable the output
```

This will switch the output ON for all time. Often a logic signal is used to turn the outputs into a high-impedance state. If, by example, we want to switch the outputs off when a state machine is idling, when idle is '1' (and idle is '0' otherwise), then we can use

```
outputs[0:2].TRST = /idle    ;enable the outputs when not idling
```

and this will disable the outputs when the idle state is entered and idle is '1'. Many PALs allow only one product term to be computed for the output enable. If a *computed enable* (one calculated to be a logic expression) requires more than a single-product term then device compilation will fail. In this case you will need to use one of the PAL outputs as an intermediate stage and use its product terms to accommodate the logic required to compute the required enable signal.

8.3.3 *Interfacing different logic technologies*

Interfacing LS or ALS TTL and CMOS logic is still a design issue, since there is still a fair amount of this TTL circuitry in use. Also, you might want to use different attributes of a particular logic technology in circuit implementation; for example, you might want to implement the majority of the switching circuitry in CMOS to use its low power consumption, whereas you might require TTL to drive a highly capacitive output stage. Although HCT CMOS is designed to be directly compatible with TTL, the lower-power version, HC, actually requires more care when interfacing to TTL circuitry. When HC CMOS drives an LS TTL load, no additional interfacing is required, since the outputs provided by the HC gate are actually directly compatible with the inputs required by the TTL gate. When TTL is used to drive a HC input, then a pull-up resistor is used to match the voltage produced by the TTL gate to the voltage required by the HC input. We then have a TTL gate driving an HC input as

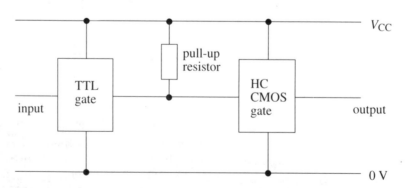

When the TTL output is HIGH, current is sourced by the TTL gate through the diode into the HC input. In this state the pull-up resistor forces the output TTL levels to reach those acceptable to the HC input since $V_{IH}(min)$ for CMOS can be greater than $V_{OH}(min)$ generated by TTL. A far easier method of interfacing HC and TTL logic is to use an intermediate HCT stage, since HCT is directly compatible with both the HC and the TTL families.

Interfacing logic families can be more complex than indicated here, particularly for fast logic and ECL. Difficulty with ECL can be exacerbated by power supply incompatibility. High switching speeds also incur problems. This requires study of *transmission lines* and *reflections*, which concerns terminating a logic signal with a correctly chosen impedance so that the logic signal is not reflected back from the input (or even just absorbed by it). Transmission lines and their characteristic impedance are vital to the correct implementation of ECL circuitry, owing to its high-speed operation. Implementing ECL circuitry and interfacing it to other logic technologies is then a very specialist area and you should consult one of the texts referenced in Section 8.8.

8.4 Driving circuits

The use of tristate logic is well established in bus driving circuits. Open-collector logic can provide a useful way to drive *light-emitting diodes* (LEDs); whereas for prototype circuits LEDs can often be driven directly from a chip output, it is better to include a resistor to limit current through the diode. There are circuits that provide LED and lamp drivers. There are also segment LED displays, which can illuminate collections of segments of an alphanumeric display. There are decoder/driver circuits for these and more sophisticated driver circuits for more complex LED and *liquid-crystal* displays (LCDs).

There are also *line driver circuits*, which can provide enabled outputs with a high current capability; these are useful when distributing signals, such as through the data bus. These can be accompanied by *line receiver circuits*, which are Schmitt trigger circuits to reduce the noise associated with a received signal. If the received signal is highly contaminated by noise and dwells around the switching threshold of the inverter then the inverter output would change rapidly, potentially giving false switching. Alternatively, if it is received by a Schmitt trigger inverter, the hysteresis associated with the switching levels within the Schmitt trigger, then the noise on the incoming signal is reduced dramatically on the output.

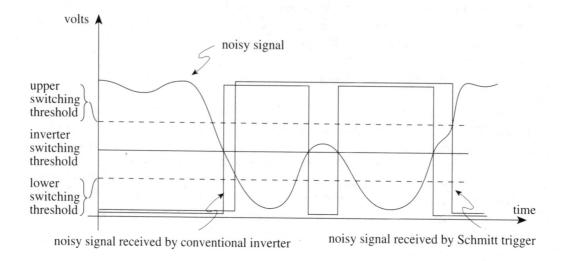

noisy signal received by conventional inverter   noisy signal received by Schmitt trigger

8.5   Clock circuits

A clock signal is usually a square-wave signal of controlled frequency. This can be provided by the output of a single inverter with the output fed back to provide the inverter input. This uses the propagation delay of the inverter to give the square-wave signal, since a LOW output is fed back to the input, which causes the output to change to HIGH T_{PLH} ns later and which then goes to LOW T_{PHL} ns later.

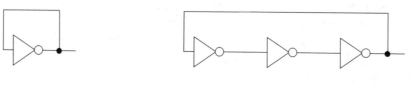

Single-inverter clock circuit   Ring inverter circuit

This can be extended to lengthen the clock period by using any odd number of inverters. It is not a very reliable circuit and, since the propagation delay is usually nanoseconds, this implies a high clock frequency. Clock circuits are sometimes called *astables*, since they are stable in neither the LOW or HIGH state, but *oscillate* between LOW and HIGH. A slower version can be achieved without an excessive number of inverters by driving the input of a Schmitt inverter with the fed-back output connected to the input by a resistor–capacitor network. This circuit actually needs the Schmitt inverter since it would not work correctly with a conventional inverter. The hysteresis in the Schmitt inverter ensures that the output oscillates with a frequency controlled by the rate at which the fed-back voltage approaches the lower switching threshold and then the upper switching threshold.

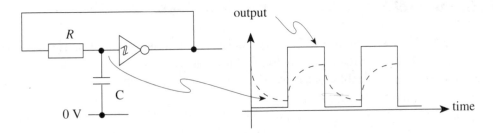

The capacitor charges and discharges to cause the input voltage to the Schmitt trigger inverter to rise and fall, according to the time constant of the resistor–capacitor network. When the inverter input exceeds the Schmitt trigger threshold, the output will switch until the input voltage passes the other switching threshold. The time constant controls the rate of rise and fall of the input voltage and hence the frequency of the output square wave. A guaranteed clock frequency is usually required in sequential circuits, and this can be achieved using a *quartz crystal oscillator*. A quartz crystal can be manufactured at relatively low cost to resonate at a fixed frequency with a specified precision. This can then be used to regulate the frequency at which a clock circuit oscillates and provides a clock signal to high accuracy. In the following circuit a 14-stage ripple counter reduces the signal from a 32 768 kHz clock signal to provide a 2 Hz output.

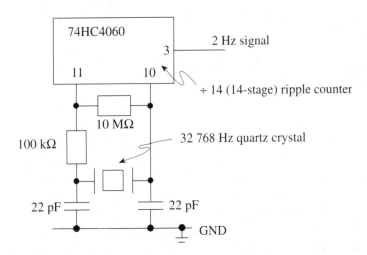

There are circuits, *clock drivers*, that are specifically designed for inclusion with a chosen quartz crystal to provide a desired clock signal, and **quartz crystal oscillator modules**, which provide a logic level clock at specified frequency without any external components.

In a synchronous sequential system the clock will be fed to the whole circuit. This implies a clock source with a high fan-out. If this is not available then it is often necessary to buffer the clock signal to ensure that power requirements are met and the circuit operates correctly.

8.6 Monostables

The *monostable* is a device aimed to provide a pulse of controlled length. It is stable in one state only, when it is not providing the pulse, and it is called a monostable for this reason. A basic circuit can use a bistable that uses the asynchronous clear input fed by a resistor–capacitor network fed by the $\overline{Q}$ output.

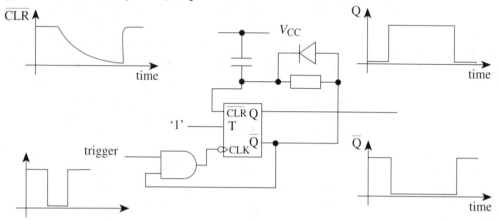

When the bistable is in its stable state, the Q output is LOW, $\overline{Q}$ is HIGH, the clear input is also HIGH and the device is perfectly stable. When the trigger signal goes LOW, the negative edge-triggered bistable output Q changes from LOW to HIGH and the $\overline{Q}$ output changes from HIGH to LOW. The capacitor then discharges through the $\overline{Q}$ output at a rate given by the time constant. The voltage at the clear input then falls consistent with the falling voltage across the capacitor. When the voltage at the clear input reaches its threshold value, the bistable is cleared, setting Q LOW and $\overline{Q}$ HIGH. The capacitor then charges up rapidly through the diode (and hence the direction of the diode connection), the clear input rises to an inactive level and the device returns to its stable state. The time between the active clock edges is often called the **trigger signal**, and the bistable returning to LOW again is determined by the time constant of the resistor–capacitor network. This then controls the length of the HIGH pulse provided at the Q output following an active clock edge. While monostables provide a pulse of specified length, the time duration of the pulse can vary with age and temperature. Capacitors leak and reduce in capacitance with age. The performance of monostables can vary with temperature. If you require a pulse of fixed and guaranteed length, you should use a quartz crystal clock and a counter to count clock cycles and hence provide a pulse of correct duration.

There are two main types of monostable. There is a *retriggerable monostable*, which provides a pulse of fixed length on receipt of every trigger signal, even when the pulse is already HIGH; there is also a *non-retriggerable* monostable, which provides one pulse of fixed length after a trigger signal, even if there are further trigger commands when the output is HIGH. The retriggerable monostable **listens** to further trigger signals when the output is active, whereas the non-retriggerable monostable **ignores** further trigger commands when the output is active. Both are available as integrated circuits. The bistable described earlier is a non-retriggerable monostable.

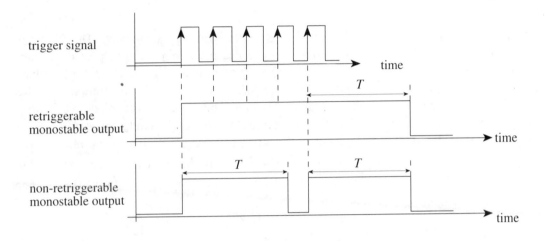

An example of an application of a retriggerable monostable is to debounce a signal derived from an optical switch. If an optical switch is used to detect the passage of some object through it, then the switch output will bounce and only settle some time after the object has passed through. The bounce can be resolved by setting the pulse length of the retriggerable monostable output to exceed greatly the expected period of the bounce in the switch output.

Though monostables are usually inadequate for providing pulses of fixed length, they can be useful when interfacing to digital circuits. The retriggerable monostable can be used to interface a push-button switch, or equivalent circuit, which provides an output that bounces and the bounce is removed by the monostable. The non-retriggerable monostable provides a pulse of fixed length at a triggering event and can be used to 'stretch' a trigger signal.

8.7 Timer chips

There are several circuits available that provide differing timing functionality. The 555 timer chip has a long pedigree in digital circuits, since it can be configured to provide circuits including astables (clocks) or monostables. Its enduring popularity is reflected by the later introduction of low-power versions. The architecture of the 555 timer chip comprises an RS bistable with inputs fed by two comparators. The comparators have a voltage derived from a potentiometer chain (as in a flash A/D converter) as the inverting input to one, comparator B (V_{TH}^-), and, as the non-inverting input to the other, comparator A (V_{TR}^+). The resistors are equal in value, so V_{TH}^- is $2V_{CC}/3$ and V_{TR}^+ is $V_{CC}/3$. The discharge input can be connected to ground through a transistor in an open-collector configuration when $\overline{Q}$ is HIGH. The architecture is then

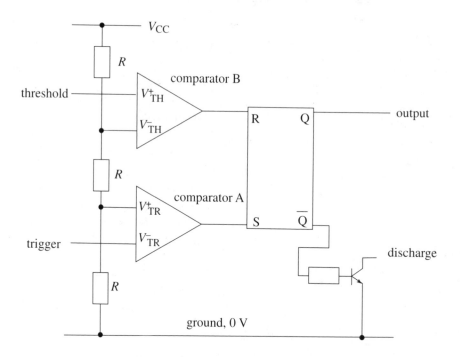

One advantage of the 555 timer chip is that relatively low clock frequencies can be achieved with capacitors of only a moderate size. Again, a resistor–capacitor network is used to control frequency of oscillation; these are connected to the inputs of the 555 chip, in particular the discharge input. When the Q output is HIGH, the $\overline{Q}$ output is LOW and the discharge transistor is OFF. The discharge input is then disconnected from ground, and the (common) voltage input fed to V_{TH}^+ and V_{TR}^- will rise as the capacitor charges. Given that the capacitor is discharged then the voltage at V_{TH}^+ and V_{TR}^- will be LOW, so the outputs of comparators A and B will be HIGH and LOW respectively, consistent with the bistable being set. The capacitor will then charge up, causing the input voltage to rise and, when V_{TR}^-

just exceeds V_{TR}^+, the output of comparator A will go LOW, but the bistable output does not change since it just changes from the set state to a latch state. When the voltage at V_{TH}^+ rises to just exceed V_{TH}^- then the output of comparator B will be HIGH, causing the bistable to reset, and the Q output changes to LOW. The discharge transistor is then switched ON through $\overline{Q}$, which is now HIGH, and the capacitor discharges. The output will remain LOW until the V_{TR}^- decreases to be just less than the V_{TR}^+ reference voltage, when the output of comparator A goes HIGH and the bistable is set again.

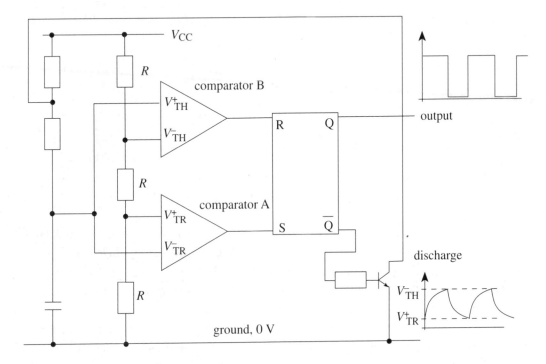

In this manner the output will SET and RESET according to the rate at which the capacitor charges and discharges, and so the external resistors and capacitor control the frequency at which the output oscillates.

The 555 timer can also be connected as a monostable by using the input to comparator A as an active low trigger input with a resistor–capacitor network attached to the input of the other comparator.

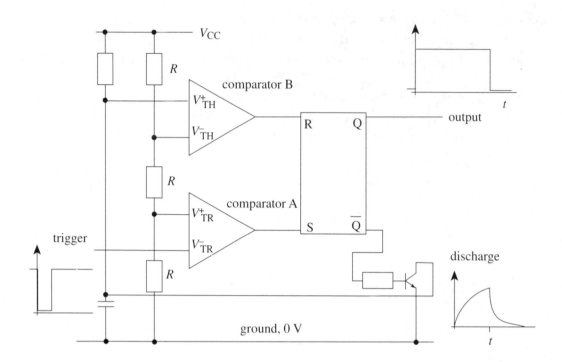

In the quiescent state, the outputs of comparators A and B will be LOW and HIGH respectively, and the bistable is RESET. When an active low trigger input causes V_{TR}^- to be less than the reference $_{TR}^+$ then comparator A output will be HIGH and the output Q will be '1'. The discharge transistor will then be OFF, so the capacitor will charge up through the external resistor and V_{TH}^+ will rise. When the trigger input V_{TR}^- returns HIGH, the bistable will latch the set state, but when V_{TH}^+ rises to just exceed V_{TH}^- then the output of comparator A will be HIGH, resetting the bistable, and the monostable output will return to '0'. The rate at which the capacitor charges controls the length of time which the output is at '1', and thus the length of the pulse output from the monostable. This is a non-retriggerable monostable.

8.8 Concluding comments and further reading

This chapter has reviewed some of the more practical aspects of digital systems design; how we connect circuits together and how we provide clock signals. There are many other aspects to such study, but which require much more detailed electronic circuit analysis. These include minimising *cross-talk*, where we reduce the possibility that a signal on one wire can corrupt, or interfere with, the signal on another. Even the layout of the power supplies on the circuit can be vital to correct circuit operation, and many circuits use a *ground plane*, which is a large area of conducting material connected to ground, 0 V, and placed underneath the switching circuitry to minimise the effects of noise: Thornton, E., *Electrical Interference and Protection* (Ellis Horwood, 1990) gives an introduction. Among manufacturers' handbooks, try the *Interface Databook* provided by manufacturers such as

National Semiconductor or Texas Instruments. Wilkinson, B. and Makki, R., *Digital System Design* (Prentice-Hall, 2nd edn, 1992) provides an introduction to transmission lines in the context of digital circuit design. Few books address circuit production and layout alone, and many provide an isolated section within a book. Catt, I., Walton, D. and Davidson, M., *Digital Hardware Design* (Macmillan, 1979) is dedicated to this subject but is perhaps a little dated now.

Testability is germane to modern production. This concerns designing circuits to ensure that their performance can be guaranteed. For further study in testability, Wilkins, B. R., *Testing Digital Circuits* (Van Nostrand-Reinhold, 1986) and Abromovici, M., Breurer, M. A. and Friedman, A. D., *Digital Systems Testing and Testable Design* (Computer Science Press, 1990) are an excellent starting point.

Appendix 1 Drawing logic circuits

A1.1 Mixed-logic convention

The book has used positive logic throughout, which is the convention that TRUE is represented by a HIGH logic level. There is also a *negative logic* convention, where TRUE is represented by a LOW logic level. Without a clear specification, the circuit logic levels can be interpreted ambiguously. Consider for example the CLEAR input to a bistable, which is often labelled as $\overline{\text{CLR}}$. This is usually LOW to clear a bistable, but does this mean that $\overline{\text{CLR}} = \text{L}$ or that CLR = L? Mixed logic is a standard which allows us to use positive and negative logic when drawing logic circuits which allow them to be interpreted unambiguously.

In mixed logic we label logic variables, which are the signals output from or input to a circuit, and we label them according to the value for TRUE. If TRUE = LOW then we can append a '.L' to the signal name. If TRUE = HIGH then we can append a '.H' to a variable. We have then specified the voltage level for any point on a circuit both as a variable and as a signal. Consider for example the following circuit for a multiplexer drawn for positive logic:

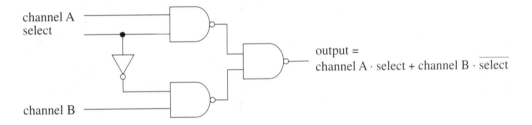

This does not give a clear specification of what levels are associated with each signal. If it is drawn using the mixed-logic convention then it becomes

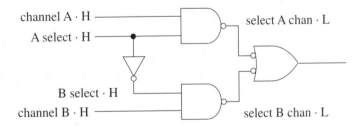

This then shows that if either of the input data is HIGH then the output is HIGH according to the level of the control signal which is HIGH to select A and LOW to select B. This is then interpreted unambiguously. In this circuit the NAND gate is replaced with an OR gate

261

with inverters at the inputs. The inverted output of the first NAND gate then connects with an inverted input to an OR gate. This is consistent with the levels specified for the signals in the circuit. In positive and negative logic there is only one way to draw a gate, and that is how it is usually realised. The mixed-logic convention actually expands the number of ways we can obtain a function and the duality between them is given by de Morgan's law. The possible circuits for OR are then

The value of the mixed-logic convention becomes much more apparent for large circuits, which may extend over a number of pages. Here the levels for any interconnecting signals need to be specified exactly, so that anyone interpreting the circuits function does so correctly and unambiguously.

The advantages of the mixed-logic convention still apply to programmable logic. However, since these are specified in software the detail of the implementation is explicit in the software. The mixed-logic convention is advantageous where we have large-scale systems containing may PLDs, and the whole system needs to be described without the extra complexity incurred by including the software specifications.

A1.2 IEEE/ANSI Logic Symbol Standard

The logic symbols used throughout this text have a distinctive shape associated with the logic function they implement. The IEEE/ANSI standard uses rectangles only to display logic functions, even up to complex sequential devices, and thus removes the distinctive shape. The standard is not ubiquitous; few PAL device handbooks use it, but many companies require its use as an internal standard. The rectangle depicts a logic function, which is written depicted inside it. The inputs and outputs are clearly labelled and circles are used to depict inverted outputs.

The IEEE/ANSI symbols for AND, OR, NAND, NOR and NOT have been given earlier, on page 17. The use of the ampersand, &, to represent AND is natural, the use of ≥ 1 to depict OR arises from the OR function being TRUE if at least more than one input is TRUE. The 1 representing NOT actually symbolises logic identity (no operation) and the inversion is actually depicted by the circle on the output. Circles on the output are again used to depict NAND and NOR. The symbols for EXCLUSIVE NOR and EXCLUSIVE OR use the fact that the output of an EXOR gate is TRUE if only one input is TRUE, depicted by the =1 in its symbol:

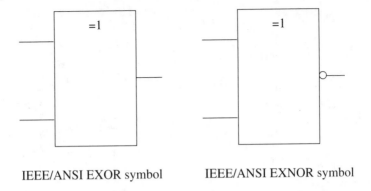

IEEE/ANSI EXOR symbol   IEEE/ANSI EXNOR symbol

These extend to more complex functions and the decoder and multiplexer:

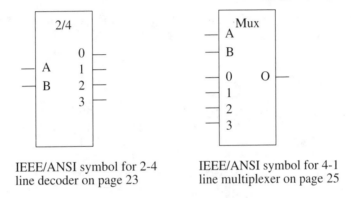

IEEE/ANSI symbol for 2-4   IEEE/ANSI symbol for 4-1
line decoder on page 23   line multiplexer on page 25

The standard extends to bistables where the system used earlier to depict clock action is the same. The clock is usually depicted alongside the circuit inputs. A D-type bistable is then

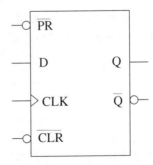

This system then extends to complex sequential logic devices. The system also extend to PAL architectures, though these are often not used in device handbooks. They do, however, provide a consistent standard for system documentation.

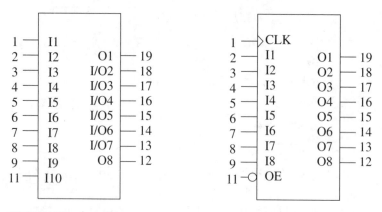

IEEE/ANSI symbol for PAL16L8   IEEE/ANSI symbol for PAL16R8

Pin numbers are included on these diagrams since they now refer to real, physical, devices. The IEEE/ANSI symbol for a PALCE16CV8 is

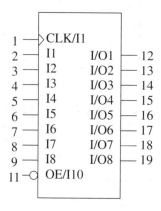

Pin 1 can actually be specified as either a logic input for combinatorial implementations or as a clock input for sequential ones and should be specified as such in circuit diagrams, omitting the clock dynamic indicator as appropriate. Pin 11 serves as an output enable for all bistables, or as an input to the device, and should be labelled appropriately in application.

Appendix 2 The PLPL programming language and the PEEL18CV8

LPL was a programming language originally introduced by, but no longer supported by, AMD Inc., though a number of institutions continue to use it. There is actually an option within the PALASM system to allow you to translate PLPL designs into PALASM. Because of PLPL's popularity, it is used to repeat the examples shown earlier in Chapter 5 with virtue of demonstrating their implementation in another language, albeit rather similar to PALASM. It is used here to introduce another PAL, the PEEL18CV8, which was originally introduced as the first electrically reprogrammable PAL in 1989. Its architecture allows for more complexity than the AMD PALCE16V8 introduced earlier.

A2.1 The PEEL 18CV8 PAL

The PEEL18CV8 is an electrically erasable PAL (it actually incorporates electrically erasable ROM technology) which has configurable outputs that can be combinational or sequential. The PEEL18CV8 has one input pin that can be used as a logic input or as a clock for sequential architectures. Another nine input pins can be used solely as logical inputs, and there are eight pins that can be used for input or output, termed *I/O pins*. The clock/logic input pin plus the nine fixed inputs and the eight I/O pins total the eighteen possible inputs. The circuit arrangement for each I/O pin is

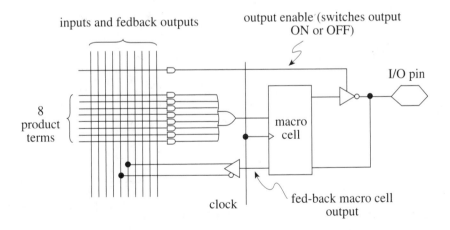

The inputs can be used to form eight product terms fed to eight *macro cells*, which drive an output pin that can be configured as a combinational or sequential output. Output pins are bidirectional. The output is actually fed through a buffer, and we enable the buffer to allow the signal onto the I/O pin. This arrangement is used for each of the eight I/O pins.

Each I/O pin can be programmed to use one of twelve different configurations. Each output is then described by a macro cell and the programmer (designer) specifies the content of the macro cell. The configuration uses multiplexers to allow different configurations to be implemented. Multiplexer A controls whether the output is purely combinational or derived from a bistable output. Multiplexer B can be enabled to switch the output on or off and controls whether the output is active high or active low. The outputs can be fed to the other circuits within the chip to allow cascading to implement larger functions or to give the next state in sequential systems; this is controlled by multiplexer C. The macro cell then implements the following circuit:

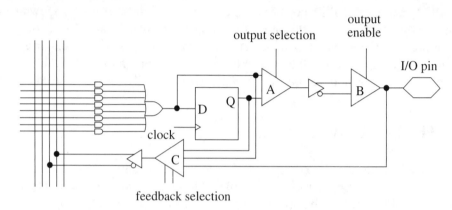

The available configurations that we can specify to be used for the macro cells are then

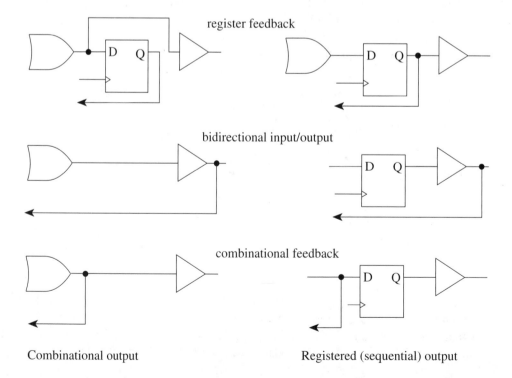

These are all for active high outputs. The first two circuits provide register feedback which allows the bistable output to be used within next-state logic for the bistables within the circuit. The output can either be derived directly from the bistable (and thus remain constant for until the next clock edge), or the output can be combinational, taken from the input to the bistable. The feedback can also allow a pin to be bidirectional and used as input or output, according to whether the output is enabled or not. When the output is enabled the circuitry attached to the output pin should be designed to expect an output signal. When the output is not enabled (and switched off) then the output can become an input and the circuitry attached to it may provide an input signal. Bidirectional I/O can use a registered or a combinational output. The feedback might also be required to be combinational, say to allow cascading inputs and hence larger logic functions in next-state circuitry feeding the bistables. With combinational feedback the output can again be either combinational or registered.

We can also specify that the outputs are active low, in which case the buffer driving the output becomes an inverting buffer. The options we can choose for an active low output are then

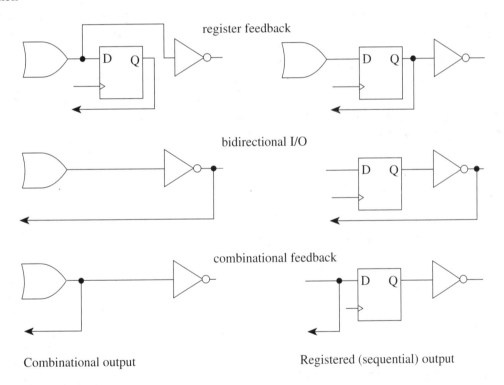

register feedback

bidirectional I/O

combinational feedback

Combinational output   Registered (sequential) output

The full device architecture then has macro cells on eight pins, making it a highly versatile device that can implement a wide range of logic functions. Ten pins can be used as input only and with the I/O pins this gives the 18 possible inputs the PEEL18CV8 can accept. One (pin 1) is used as the clock input if sequential circuitry is selected.

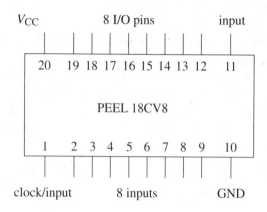

A2.2 The PLPL programming language

The PLPL language is again a specification language and **not** a programming language proper. All instructions in a PLPL specification are 'executed' at the same time.

The **keywords** in PLPL are

BEGIN	to bracket sets of instructions and to start the program and test vectors
CASE	to choose from an ordered list addressed by the CASE statement
DEFINE	to label statements for shorthand use
DEVICE	to specify the device to be programmed
ELSE	in IF, execute the following instruction(s) if a condition has not been met
ENABLE	switch something on (usually the outputs)
END	goes with BEGIN
IF	if some condition is met then execute the following instruction(s)
PIN	label a pin for use within the program
PRESET	used to preload the device
RESET	used to clear the device
THEN	goes with IF, execute following instructions(s) if condition is met

The **operators** are

+	logical OR
*	logical AND
%	exclusive OR operator
( )	used to surround logic expressions
/	logical inversion, NOT
=	assignment operator
#b	binary
#d	decimal
#h	hexadecimal

The **punctuation marks** are

" "   used to enclose comments
;   used to terminate statements
:   used to indicate a range of values
'   this is used to concatenate values and variables in CASE statements
.   indicates the end of the file (follows the END statement linked with the program
      BEGIN at the start of the program

The **pin specifications** are not truly a part of PLPL but are part of the parts description database file. They specify how you can use the pins on a chip.

```
pin=input combinatorial
```
active high logic input

```
/pin=input combinatorial
```
active low logic input

```
pin=output com
```
active high logic output which is also fed back

```
/pin=output com
```
active low logic output which is also fed back

```
pin=output registered active_high
```
active high output derived from a bistable output and is also fed back

```
pin=output registered active_low
```
active low output derived from a bistable output and is also fed back

```
pin=output registered active_high feed_reg
```
active high output derived from a bistable output, bistable output fed back

```
pin=output registered active_low feed_reg
```
active low output derived from a bistable output, bistable output fed back

```
pin=output registered active_high feed_com
```
active high output derived from a bistable output, bistable input fed back

```
pin=output registered active_low feed_com
```
active low output derived from a bistable output, bistable input fed back

Use of the commands, operators and punctuation marks will be used in design examples here to clarify their use. The specification is contained in a file with extension .PLD, e.g. WONDER.PLD, which is created using an editor. The PLD specification is then compiled to produce a fuse map file, `*.JED`. We again introduce test vectors which allow us to check that the device's operation (as worked out in the fuse map) is consistent with our expectations on it. We will now work through several of the earlier examples again, implementing them in PLPL for a PEEL18CV8.

A2.3 Examples in PLPL and PEEL18CV8

Wonderland's alarm

We shall now implement the alarm circuit using a PEEL18CV8, a replica of Section 5.1 but using the PLPL language and system. The specification in a PLPL description is contained in a file WONDER.PLD.

```
DEVICE wonderland_alarm (p18cv8)       "Name the PEEL18CV8"

"PAL description to provide John Wonderland's alarm"
PIN      "Now label the pins as inputs and outputs"
         alice  =  2 (input combinatorial)          "Alice =1=inside"
         bo     =  3 (input combinatorial)          "Bo =1=inside"
         chas   =  4 (input combinatorial)          "Chas =1=inside"
         dave   =  5 (input combinatorial)          "Dave =1=inside"
         alarm  = 16 (output com active_high)       "alarm =1=on"
;

BEGIN "Now determine the logical relationship between input and output"
ENABLE (alarm);                 "enable the alarm to be on or off"
IF ((alice*/bo*(/chas+/dave))+   "If (bo is outside with dave or
                                  chas", but not alice) or"
   (/alice*bo*(chas+dave)))       "bo is inside with dave or chas only"
      THEN                                "then"
          alarm=1;                          "ring the alarm"
      ELSE                            "otherwise"
          alarm=0;                          "the alarm is off"
END.
```

The PLPL test vectors are appended to the specification. These are:

```
TEST_VECTORS
IN  alice,bo,chas,dave;   "Define the inputs to be tested"
OUT  alarm;               "and then the output"
```

```
BEGIN
"then define the expected output state (High, H, or Low, L)"
"for each input combination, starting with inputs specified as 0 or 1"
"alice   bo     chas    dave    alarm"
"-------------------------------------"
0        0      0       0       L;   "all outside"
0        0      0       1       L;   "dave outside alone"
0        0      1       0       L;   "chas outside alone"
0        0      1       1       L;   "chas and dave outside"
0        1      0       0       L;   "bo inside alone"
0        1      0       1       H;   "bo inside with dave"
0        1      1       0       H;   "bo inside with chas"
0        1      1       1       H;   "bo inside with chas and dave"
1        0      0       0       H;   "bo outside with chas and dave"
1        0      0       1       H;   "bo outside with chas"
1        0      1       0       H;   "bo outside with dave"
1        0      1       1       L;   "bo outside alone"
1        1      0       0       L;   "alice chaperoning bo"
1        1      0       1       L;   "     "        "
1        1      1       0       L;   "     "        "
1        1      1       1       L;   "     "        "
END.
```

We then compile the PLPL program to work out a fuse plot. It first works out the minterms that implement the design and then minimises these to the prime implicants, which provide a minimal implementation in an AND/OR configuration. The intermediate file WONDER.INT contains the specification showing which terms switch the output to '1' (appended with #1) and those which force the output to be '1' (appended with #0).

```
wonderland_alarm
p18cv8
alice 2 INPUT COMBINATORIAL @
bo 3 INPUT COMBINATORIAL @
chas 4 INPUT COMBINATORIAL @
dave 5 INPUT COMBINATORIAL @
alarm 16 OUTPUT COM ACTIVE_HIGH @
16 = /4*2*/3*#1 + /5*2*/3*#1 + 4*/2*3*#1 + 5*/2*3*#1 + /2*/3*#0 +
     2*3*#0 + 4*2*5*#0 + /4*/2*/5*#0;
16.enable = #1;
```

This is then minimised to produce the WONDER.LST file, which contains the minimised equations for implementation together with a diagram showing the chip connections.

```
Boolean Equations from Design [wonderland_alarm]
alarm(H) = /chas*alice*/bo +
  alice*/bo*/dave +
  chas*/alice*bo +
  /alice*bo*dave ;
alarm.enable = #1 ;


-> [p18cv8] Chip Diagram for Design [wonderland_alarm]


                        ------v------
           *    [>>--| 1      20 |--<Vcc   *
        alice   [>>--| 2      19 |--[<>]   *
           bo   [>>--| 3      18 |--[<>]   *
         chas   [>>--| 4      17 |--[<>]   *
         dave   [>>--| 5      16 |--[>>   alarm
           *    [>>--| 6      15 |--[<>]   *
           *    [>>--| 7      14 |--[<>]   *
           *    [>>--| 8      13 |--[<>]   *
           *    [>>--| 9      12 |--[<>]   *
           *   Gnd>--| 10     11 |--<<]    *
                     |_____| Note: * = no name
```

```
      ! ---- Device Usage ---- !
[            Pin                || Product Terms (PTs)]
[       Name        #     Type  ||  Available | Used ]
[   ** Unnamed **]  [ 1]  [Clk/Inp] ||  [ 0]   |      ]
[        alice]     [ 2]  [ Input]  ||  [ 0]   |      ]
[           bo]     [ 3]  [ Input]  ||  [ 0]   |      ]
[         chas]     [ 4]  [ Input]  ||  [ 0]   |      ]
[         dave]     [ 5]  [ Input]  ||  [ 0]   |      ]
[   ** Unnamed **]  [ 6]  [ Input]  ||  [ 0]   |      ]
[   ** Unnamed **]  [ 7]  [ Input]  ||  [ 0]   |      ]
[   ** Unnamed **]  [ 8]  [ Input]  ||  [ 0]   |      ]
[   ** Unnamed **]  [ 9]  [ Input]  ||  [ 0]   |      ]
[   ** Unnamed **]  [ 10] [  GND]   ||  [ 0]   |      ]
[   ** Unnamed **]  [ 11] [ Input]  ||  [ 0]   |      ]
[   ** Unnamed **]  [ 12] [  I/O]   ||  [ 8]   |      ]
[   ** Unnamed **]  [ 13] [  I/O]   ||  [ 8]   |      ]
[   ** Unnamed **]  [ 14] [  I/O]   ||  [ 8]   |      ]
[   ** Unnamed **]  [ 15] [  I/O]   ||  [ 8]   |      ]
[        alarm]     [ 16] [ Output] ||  [ 8]   | [ 4] ( 50.0%)
[   ** Unnamed **]  [ 17] [  I/O]   ||  [ 8]   |      ]
[   ** Unnamed **]  [ 18] [  I/O]   ||  [ 8]   |      ]
[   ** Unnamed **]  [ 19] [  I/O]   ||  [ 8]   |      ]
[   ** Unnamed **]  [ 20] [  Vcc]   ||  [ 0]   |      ]
      << Device Usage >>
- Available PTs = [  64] - PTs Used = [  4] - Utilization = [  6.3%]
```

The JEDEC file, `WONDER.JED` again shows which fuses are blown. We then simulate the device to test the functionality of the program. We use a simulator that will extract the test vectors specified in the original program and simulate the PLD with them. It does not evaluate the program output for each input combination. A synthesised version of the 18CV8 (specified from a database file) is developed from the JEDEC file and the test vectors are applied to the tailored logic simulator so created. The outputs from the synthesised 18CV8 are then compared with those specified in the test vector set. Should the expected output disagree with the H or L we have specified in the test vector set then the simulator program will mark it as an error in the result and end with a total of these errors.

```
Simulating [wonderland.jed] (device file [.\s18cv8])
  with vectors in file [wonderland.tst]
==> Device Pin #: [0000 0000 0111 1111 1112]    Device Pin #: [0000 0000 0111 1111 1112]
- - - - - - - - [1234 5678 9012 3456 7890]                   [1234 5678 9012 3456 7890]
- Applying [  1]: [X000 0XXX XNXX XXXL XXXN] - Applying [  9]: [X100 0XXX XNXX XXXH XXXN]
  Calculated  => [X000 0XXX XNXZ ZZZL ZZZN]   Calculated  => [X100 0XXX XNXZ ZZZH ZZZN]
- Applying [  2]: [X000 1XXX XNXX XXXL XXXN] - Applying [ 10]: [X100 1XXX XNXX XXXH XXXN]
  Calculated  => [X000 1XXX XNXZ ZZZL ZZZN]   Calculated  => [X100 1XXX XNXZ ZZZH ZZZN]
- Applying [  3]: [X001 0XXX XNXX XXXL XXXN] - Applying [ 11]: [X101 0XXX XNXX XXXH XXXN]
  Calculated  => [X001 0XXX XNXZ ZZZL ZZZN]   Calculated  => [X101 0XXX XNXZ ZZZH ZZZN]
- Applying [  4]: [X001 1XXX XNXX XXXL XXXN] - Applying [ 12]: [X101 1XXX XNXX XXXL XXXN]
  Calculated  => [X001 1XXX XNXZ ZZZL ZZZN]   Calculated  => [X101 1XXX XNXZ ZZZL ZZZN]
- Applying [  5]: [X010 0XXX XNXX XXXL XXXN] - Applying [ 13]: [X110 0XXX XNXX XXXL XXXN]
  Calculated  => [X010 0XXX XNXZ ZZZL ZZZN]   Calculated  => [X110 0XXX XNXZ ZZZL ZZZN]
- Applying [  6]: [X010 1XXX XNXX XXXH XXXN] - Applying [ 14]: [X110 1XXX XNXX XXXL XXXN]
  Calculated  => [X010 1XXX XNXZ ZZZH ZZZN]   Calculated  => [X110 1XXX XNXZ ZZZL ZZZN]
- Applying [  7]: [X011 0XXX XNXX XXXH XXXN] - Applying [ 15]: [X111 0XXX XNXX XXXL XXXN]
  Calculated  => [X011 0XXX XNXZ ZZZH ZZZN]   Calculated  => [X111 0XXX XNXZ ZZZL ZZZN]
- Applying [  8]: [X011 1XXX XNXX XXXH XXXN] - Applying [ 16]: [X111 1XXX XNXX XXXL XXXN]
  Calculated  => [X011 1XXX XNXZ ZZZH ZZZN]   Calculated  => [X111 1XXX XNXZ ZZZL ZZZN]
                              Simulation Completed: Errors [  0]
```

Synchronous modulo-8 counter

This is the same as the counter in Section 5.5.2 using a PLPL description, which is

```
DEVICE counter (p18cv8)   "Synchronous Modulo 8 Counter"
"First define the pins"
PIN
        clk   = 1 (clock)      "clock input"
        z[2:0] =  12,13,14 (output reg active_high feed_reg);
     "registered outputs to remember the states"
BEGIN
   ENABLE (z[2:0]);  "enable the state outputs"
   CASE(z[2:0])
```

```
    BEGIN
        #b000)  z[2:0]=#b001;  "the next state after state 0 is 1"
        #b001)  z[2:0]=#b010;  "from 1 we go to 2"
        #b010)  z[2:0]=#b011;  "and from 2 to 3"
        #b011)  z[2:0]=#b100;  "3 to 4"
        #b100)  z[2:0]=#b101;  "4 to 5"
        #b101)  z[2:0]=#b110;  "5 to 6"
        #b110)  z[2:0]=#b111;  "6 to 7"
        #b111)  z[2:0]=#b000;  "and from 7 back to the start"
    END;
END.
```

The test vectors then introduce values for the clock; these can be c to denote a clock event or p to denote presetting the device (labelling the device's initial state).

```
TEST_VECTORS
IN  clk; "There are no combinational inputs in the design"
OUT  z[2:0]; "These are the three state outputs"
BEGIN
"clk z[2:0] "
"-----------"
p    LLL;   "start the test vectors from 000"
c    LLH;   "execute sequence, at the first clock move from 000 to 001"
c    LHL;   "at the second clock move from 001 to 010 (the next state)"
c    LHH;   "and then from 010 to 011"
c    HLL;   "then to 100"
c    HLH;   "and on to 101"
c    HHL;   "use of H or L designates the expected output value"
c    HHH;   "this is compared with that calculated from compilation"
c    LLL;   "Now check it starts again"
END.
```

The compiled equations for the counter are

```
Boolean Equations from Design [counter]

z[2](H) = /z[2]*z[1]*z[0] +
   z[2]*/z[1] +
   z[2]*/z[0] ;
z[2].enable = #1 ;
z[1](H) = /z[1]*z[0] +
   z[1]*/z[0] ;
z[1].enable = #1 ;
z[0](H) = /z[0] ;
z[0].enable = #1 ;
```

There are two stages to testing in PLPL; the first is to check that the test vectors show that the operation of the device is consistent with its design.

```
Simulating [counter.jed] (device file [.\s18cv8])
  with vectors in file [counter.tst]
```

```
==> Device Pin #: [0000 0000 0111 1111 1112]     Device Pin #: [0000 0000 0111 1111 1112]
- - - - - - - - - [1234 5678 9012 3456 7890]                  [1234 5678 9012 3456 7890]
- Applying [  1]: [PXXX XXXX XNXL LLXX XXXN]  - Applying [  6]: [CXXX XXXX XNXH LHXX XXXN]
  Calculated   => [0XXX XXXX XNXL LLZZ ZZZN]    Calculated   => [0XXX XXXX XNXH LHZZ ZZZN]
- Applying [  2]: [CXXX XXXX XNXL LHXX XXXN]  - Applying [  7]: [CXXX XXXX XNXH HLXX XXXN]
  Calculated   => [0XXX XXXX XNXL LHZZ ZZZN]    Calculated   => [0XXX XXXX XNXH HLZZ ZZZN].
- Applying [  3]: [CXXX XXXX XNXL HLXX XXXN]  - Applying [  8]: [CXXX XXXX XNXH HHXX XXXN]
  Calculated   => [0XXX XXXX XNXL HLZZ ZZZN]    Calculated   => [0XXX XXXX XNXH HHZZ ZZZN]
- Applying [  4]: [CXXX XXXX XNXL HHXX XXXN]  - Applying [  9]: [CXXX XXXX XNXL LLXX XXXN]
  Calculated   => [0XXX XXXX XNXL HHZZ ZZZN]    Calculated   => [0XXX XXXX XNXL LLZZ ZZZN]
- Applying [  5]: [CXXX XXXX XNXH LLXX XXXN]  - Simulation Completed: Errors [   0]
  Calculated   => [0XXX XXXX XNXH LLZZ ZZZN]
```

The device simulates successfully and propagates correctly through the state sequence. PLPL also provides a waveform diagram contained in a file with the extension .WAV. Those to which no output has been attached are denoted Z. A waveform is attached to pins specified as output, in this case pins 12, 13 and 14. There are no inputs to the circuit as denoted by X on the input pins. The clock is indicated with an arrow. The two power supplies are denoted N for no connection

```
  << Waveform Diagram >>
- simulation model  (JEDEC Map) [counter.jed]
- inputs/stimuli (test vectors) [counter.tst]
- device [.\s18cv8]
Symbols: Output          [‖] Input          [|] Not tested [N]
         Input Floating [F] Output Hi-Z [Z] Don't Care [X]
         Positive Clock [-»]      Negative Clock [«-]
V #  0 0 0 0  0 0 0 0  0 1 1 1  1 1 1 1  1 1 1 2  These are the pin numbers
  #  1 2 3 4  5 6 7 8  9 0 1 2  3 4 5 6  7 8 9 0
  1 | X X X ⊦X X X X ⊦X  NX ‖  ⊦|‖  ‖ Z Z ⊦Z Z Z  N⊦ Now we can see how the
  2 | X X X ⊦X X X X ⊦X  NX ‖  ⊦|‖  ‖ Z Z ⊦Z Z Z  N⊦ outputs (or states) change.
  2 ⊦»X X X ⊦X X X X ⊦X  NX ‖  ⊦|‖ ⌐Z Z ⊦Z Z Z  N⊦ Here changing from state
  3 | X X X ⊦X X X X ⊦X  NX ‖  ⊦|‖  ‖Z Z ⊦Z Z Z  N⊦ 0 to state 1.
  3 ⊦»X X X ⊦X X X X ⊦X  NX ‖  ⊦|⌐⌐Z Z ⊦Z Z Z  N⊦ And here from 001 to 010
  4 | X X X +X X X X +X  NX ‖  +  ‖ Z Z +Z Z Z  N+
  4 ⊦»X X X +X X X X +X  NX ‖  +  ‖⌐Z Z +Z Z Z  N+ Note that movement is at
  5 | X X X ⊦X X X X ⊦X  NX ‖  ⊦ | ‖ Z Z ⊦Z Z Z  N⊦ positive clock edges.
  5 ⊦»X X X ⊦X X X X ⊦X  NX ⌐⊦⌐⌐Z Z ⊦Z Z Z  N⊦
  6 | X X X ⊦X X X X ⊦X  NX ‖⊦|‖  ‖ Z Z ⊦Z Z Z  N⊦
  6 ⊦»X X X ⊦X X X X ⊦X  NX ‖⊦|‖ ⌐Z Z ⊦Z Z Z  N⊦
  7 | X X X ⊦X X X X ⊦X  NX ‖⊦|‖  ‖ Z Z ⊦Z Z Z  N⊦
```

```
7 |-»X  X  X  |-X  X  X  X  |-X  NX  || |-⌐ ⌐|Z  Z  |-Z  Z  Z  N|-
8 |  X  X  X  +X  X  X  X  +X  NX  ||+  ||  Z  Z  +Z  Z  Z  N|-
8 |-»X  X  X  +X  X  X  X  +X  NX  ||+  ||-Z  Z  +Z  Z  Z  N|-
9 |  X  X  X  |-X  X  X  X  |-X  NX  ||+  ||  Z  Z  |-Z  Z  Z  N|-
9 |-»X  X  X  |-X  X  X  X  |-X  NX  ⌐|-⌐|⌐|Z  Z  |-Z  Z  Z  N|-
```

JK bistable

One software description can use a CASE statement as

```
DEVICE JK_bistable (p18cv8)
"JK bistable made from a positive edge triggered D-type"

PIN "first define all active connections: clock, inputs and outputs"
  clk   =  1    (clock)
  j  =  2    (input combinatorial) "bistable input"
  k  =  3    (input combinatorial) "bistable input"
  q  =  12   (output reg active_high feed_reg);   "bistable output"
BEGIN
    ENABLE(q); "to permanently enable output"
"specify function using CASE statement implementing characteristic table"
    CASE(j,k)
    BEGIN
    #b00) q=q;  "latch"
    #b01) q=0;  "reset"
    #b10) q=1;  "set"
    #b11) q=/q; "toggle"
    END;
END.


TEST_VECTORS
IN clk,j,k;      "test as function of inputs and clock"
OUT q;           "testing a single output"

BEGIN
"clk  j  k  q"
"------------"
p      x  x  1 ; "preload set state"
c      1  1  L ; "test toggle"
c      1  0  H ; "test set"
c      1  0  H ; "remain set"
c      0  0  H ; "latch set"
c      0  1  L ; "test reset"
END.
```

with the result:

```
Boolean Equations from Design [jk_bistable]

q(H) = q*/k +
  /q*j ;
q.enable = #1 ;
```

The alternative description is

```
DEVICE JK_bistable (p18cv8)

"JK bistable made from a positive edge triggered D-type"
PIN "first define all active connections: clock, inputs and outputs"
  clk   =  1    (clock)
  j  =  2    (input combinatorial)  "bistable input"
  k  =  3    (input combinatorial)  "bistable input"
  q  =  12   (output reg active_high feed_reg);   "bistable output"
BEGIN
    ENABLE(q); "to permanently enable output"
"specify operation by interpreting functionality characteristic table"
    IF (j) THEN IF (k) THEN q=/q;   "toggle since j=k=1"
                       ELSE q=1;    "set since j=1, k=0"
           ELSE IF (k) THEN q=0;    "reset since j=0, k=1"
                       ELSE q=q;    "latch when j=k=0"
END.
```

Johnson counter

This is the Johnson counter defined in Section 5.6.2.

```
DEVICE johnson_counter (p18cv8)
"pld for johnson counter"


PIN    "first define all active connections: clock, inputs and outputs"
  clk  =  1    (clock)
  z[0:2]  =  12:14    (output reg active_high feed_reg)  "state outputs"
;
BEGIN
    ENABLE(z[0:2]); "to permanently enable output"
"specify function using CASE statement implementing next state"
    CASE (z[0:2])
    BEGIN
    #b000)  z[0:2]=#b100;
    #b100)  z[0:2]=#b110;
    #b110)  z[0:2]=#b111;
    #b111)  z[0:2]=#b011;
```

```
       #b011)  z[0:2]=#b001;
       #b001)  z[0:2]=#b000;
       #b101)  z[0:2]=#b000;    "reset unused states to zero"
       #b010)  z[0:2]=#b000;
       END;
END.


TEST_VECTORS

IN clk;        "test as function of inputs and clock"
OUT z[0:2];              "testing a single output"
BEGIN
"clk   z[0:2]"
"------------"
p       LLL; "preload zero state"
c       HLL;
c       HHL;
c       HHH;
c       LHH;
c       LLH;
c       LLL; "sequence complete"
p       HLH; "preset unused state"
c       LLL; "check it enters the state sequence"
c       HLL;
p       LHL; "check the other"
c       LLL;
END.
```

This then results in the compiled equations:

```
Boolean Equations from Design [johnson_counter]

z[0](H) = /z[1]*/z[2] +
  z[0]*/z[2] ;
z[0].enable = #1 ;


z[1](H) = z[0]*/z[2] +
  z[0]*z[1] ;
z[1].enable = #1 ;


z[2](H) = z[0]*z[1] +
  z[1]*z[2] ;
z[2].enable = #1 ;
```

Sequence detector

This is the same sequence detector as defined earlier in Section 5.6.3. The PLPL specification is quite similar to the PALASM specification earlier. This is the same for the data logger, so only the sequence detector is given here.

```
DEVICE sequence_detector (p18cv8)
"A device to detect and count the occurrence of the sequence 000111
 within a serial data input stream"

"First define the pins"
PIN
 clk                 =  1    (clock) "clock input"
 data                =  2    (input combinatorial) "input serial data"
 states[2:0]         = 12:14 (output reg active_high feed_reg)
                              "registered outputs to remember the states"
 sequence_detected = 15 (output com) "signal to say that sequence found"
;


BEGIN
    ENABLE (states[2:0],sequence_detected);
      "enable the output and states"
    CASE(states[2:0])
      BEGIN
         #b000) IF (/data)
                  THEN states[2:0]=#b001; "find first 0"
                  ELSE states[2:0]=#b000; "1 detected, wait for first 0"
         #b001) IF (/data)
                  THEN states[2:0]=#b010; "data = second 0"
                  ELSE states[2:0]=#b000; "data=1 so restart"
         #b010) IF (/data)
                  THEN states[2:0]=#b011; "data = third 0"
                  ELSE states[2:0]=#b000; "data=1 so restart"
         #b011) IF (data)
                  THEN states[2:0]=#b100; "find first 1"
                  ELSE states[2:0]=#b011; "find another 0 so restart"
         #b100) IF (data)
                  THEN states[2:0]=#b101; "data = second 1"
                  ELSE states[2:0]=#b001; "data=0 so restart at 001"
         #b101) IF (data)
                  THEN states[2:0]=#b110; "data = third 1"
                  ELSE states[2:0]=#b001; "found 0 so start there"
         #b110) BEGIN
                  sequence_detected=1; "output sequence detected"
```

```
                        IF (data)
                        THEN states[2:0]=#b000;
            "restart from beginning since 1 detected"
                            ELSE states[2:0]=#b001; "data=0 so restart at 001"
                        END;
            END;
    END.


TEST_VECTORS

IN   clk,data; "use clock and data presented"
OUT   states[2:0], sequence_detected;

BEGIN
"clk   data    states[2:0]   sequence_detected"
"----------------------------------------"
p       x       LLL               L;
c       0       LLH               L;
c       0       LHL               L;
c       0       LHH               L;
c       1       HLL               L;
c       1       HLH               L;
c       1       HHL               H;
c       1       LLL               L;
c       0       LLH               L;
c       1       LLL               L;
c       0       LLH               L;
c       0       LHL               L;
c       0       LHH               L;
c       1       HLL               L;
c       0       LLH               L;

END.
```

Data logger

This is a repeat of the design in Section 5.6.4, using PLPL.

```
DEVICE data_logger (p18cv8)
"A device to detect and count the occurrence of 000 then 111
 in a serial data input stream"
 "uses grey code state allocation for minimisation"
```

```
"First define the pins"
PIN
        clk    =  1 (clock)        "clock input"
        data   =  2 (input combinatorial) "input serial data"
        start  =  3 (input combinatorial)
        start0 =  4 (input combinatorial)
        start1 =  5 (input combinatorial)
        states[3:0] = 12:15 (output reg active_high feed_reg)
                            "registered outputs to remember the states"
        count[2:0] = 16:18 (output reg active_high feed_reg)
        ;                   "counter outputs"
BEGIN
    ENABLE (states[3:0], count[2:0]);  "enable the state outputs"
    CASE(states[3:0])
     BEGIN
     #b0000) BEGIN
             IF (start)
                 THEN states[3:0]=#b0100; "check for 3 0s"
                 ELSE states[3:0]=#b0000; "wait for start"
             count[2:0]=count[2:0];
             END;
     #b0100) BEGIN
             IF (start0*/data*/start1)
                 THEN states[3:0]=#b0110; "data = first 0"
                 ELSE states[3:0]=#b0100; "data=1 or start0=0 so restart"
             count[2:0]=count[2:0];
             END;
     #b0110) BEGIN
             IF (start0*/data*/start1)
                 THEN states[3:0]=#b0111; "data = second 0"
                 ELSE states[3:0]=#b0100; "data=1 or start0=0 so restart"
             count[2:0]=count[2:0];
             END;
     #b0111) BEGIN
             IF (start0*/data*/start1)
                 THEN states[3:0]=#b0101; "data = third 0"
                 ELSE states[3:0]=#b0100; "data=1 or start0=0 so restart"
             count[2:0]=count[2:0];
             END;
     #b0101) BEGIN
             IF (start0*start1)
                 THEN states[3:0]=#b1101; "check for 3 1s after start1"
                 ELSE states[3:0]=#b0000; "both starts not 1 so reset"
             count[2:0]=count[2:0];
             END;
     #b1101) BEGIN
             IF (start1*data*/start0)
                 THEN states[3:0]=#b1111; "data = first 1"
                 ELSE states[3:0]=#b1101; "data=0 or start1=0 so restart"
```

```
                    count[2:0]=count[2:0];
                END;
        #b1111) BEGIN
                IF (start1*data*/start0)
                    THEN states[3:0]=#b1110; "data = second 1"
                    ELSE states[3:0]=#b1101; "data=0 or start1=0 so restart"
                count[2:0]=count[2:0];
                END;
        #b1110) BEGIN
                IF (start1*data*/start0)
                    THEN states[3:0]=#b1100; "data = third 1"
                    ELSE states[3:0]=#b1101; "data=0 or start1=0 so restart"
                count[2:0]=count[2:0];
                END;
        #b1100) IF (start1)
                    THEN
                        BEGIN
                        CASE (count[2:0])   "now decrement the counter"
                                BEGIN
                                #b111) count[2:0]=#b110;
                                #b110) count[2:0]=#b101;
                                #b101) count[2:0]=#b100;
                                #b100) count[2:0]=#b011;
                                #b011) count[2:0]=#b010;
                                #b010) count[2:0]=#b001;
                                #b001) count[2:0]=#b000;
                                #b000) count[2:0]=#b111;
                                END;
                            IF (start) THEN states[3:0]=#b0100; "restart with 0s"
                        END;
                    ELSE states[3:0]=#b0000; "wait for restart"
        #b0001,
        #b0010,
        #b0011,
        #b1000,    "for the unused states, check the start signal first and"
        #b1001,    "if the 0s are starting, go to that state"
        #b1010,
        #b1011) BEGIN
                IF (start)
                    THEN states[3:0]=#b0100;
                    ELSE states[3:0]=#b0000;
                count[2:0]=count[2:0];
                END;
        END;
END.


TEST_VECTORS
IN  clk,data,start,start0,start1; "use clock and data presented"
OUT  states[3:0], count[2:0];
```

```
BEGIN
"clk  data    start   start0   start1   states[3:0]   count[2:0]"
"-----------------------------------------------------------------"
p      x       x       x        x        LLLL          LLL;
c      x       0       x        x        LLLL          LLL;
c      x       1       x        x        LHLL          LLL;
c      0       x       1        0        LHHL          LLL;
c      0       x       1        0        LHHH          LLL;
c      0       x       1        0        LHLH          LLL;
c      x       x       1        1        HHLH          LLL;
c      1       x       0        1        HHHH          LLL;
c      1       x       0        1        HHHL          LLL;
c      1       x       0        1        HHLL          LLL;
c      x       1       x        1        LHLL          LLH;
c      0       x       1        0        LHHL          LLH;
c      1       x       0        x        LHLL          LLH;
c      0       x       1        0        LHHL          LLH;
c      0       x       1        0        LHHH          LLH;
c      0       x       1        0        LHLH          LLH;
c      x       x       1        1        HHLH          LLH;
c      0       x       1        x        HHLH          LLH;
END.
```

Finally, we check the test vectors to see that the device operates correctly. The waveform listing is

```
<< Waveform Diagram >>
- simulation model   (JEDEC Map) [datalo2.jed]
- inputs/stimuli (test vectors) [datalo2.tst]
- device [c:\plpl\s18cv8]
Symbols: Output          [‖] Input          [|] Not tested [N]
         Input Floating [F] Output Hi-Z [Z] Don't Care [X]
         Positive Clock [-»]     Negative Clock [«-]
V #  0 0 0 0  0 0 0 0  0 1 1 1  1 1 1 1  1 1 1 2
  #  1 2 3 4  5 6 7 8  9 0 1 2  3 4 5 6  7 8 9 0
  1    X X X -X X X X -X  NX ‖        ‖      Z  N
  2    X | X -X X X X -X  NX ‖        ‖      Z  N
  2  -»X | X -X X X X -X  NX ‖        ‖      Z  N
  3    X ⌐X -X X X X -X  NX ‖        ‖      Z  N
  3  -»X  |X -X X X X -X  NX ‖        ‖      Z  N
  4      X   -|    X X X -X  NX ‖      ‖      Z  N
  4  -»   X   -|    X X X -X  NX ‖      ‖      Z  N
  5      X        X X X -X  NX ‖      ‖      Z  N
  5  -»   X        X X X -X  NX ‖      ‖      Z  N
  6      X        X X X -X  NX ‖      ‖      Z  N
  6  -»   X        X X X -X  NX ‖      ‖      Z  N
  7    X X        X X X -X  NX ‖      ‖      Z  N
  7  -»X X        -      X X X -X  NX ‖      ‖      Z  N
  8     |X   ⌐      X X X -X  NX ‖      ‖      Z  N
  8  -»  |X   ⌐      X X X -X  NX ‖      ‖      Z  N
```

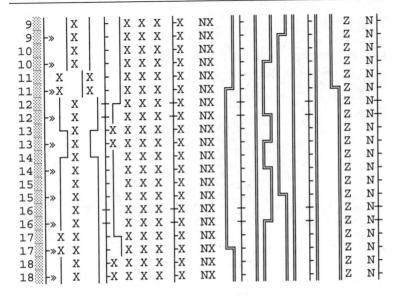

This is consistent with the required function of the device.

Successive-approximation programmable register

```
DEVICE programmable_register (p18cv8)
"2-bit programmable register for successive approximation converter"
"First define the pins"
PIN
     clk     = 1 (clock)        "clock input"
     vo_gr_vi = 2 (input combinatorial)    "1 if vout > vin"
     start   = 3 (input combinatorial)     "1 for start to convert"
     register[2:0] =  12,13,14 (output reg active_high feed_reg)
                  "registered outputs of programmable register"
;
BEGIN
   ENABLE (register[2:0]);  "enable the register outputs"
"register coding is [2:1:0] = [conversion complete:MSB:LSB]"
   CASE(register[2:0])
     BEGIN
      "first test to see whether we start or not"
        0) IF (start) THEN register[2:0]=2; "if start then set MSB"
                    ELSE register[2:0]=0; "otherwise stay at beginning"
        2) IF (vo_gr_vi) THEN register[2:0]=1; "reset MSB, set LSB"
                    ELSE register[2:0]=3; "keep MSB, set LSB"
        3) IF (vo_gr_vi) THEN register[2:0]=6; "end, MSB set LSB reset"
                    ELSE register[2:0]=7; "end, MSB and LSB set"
        1) IF (vo_gr_vi) THEN register[2:0]=4; "end, MSB and LSB reset"
                    ELSE register[2:0]=5; "end, MSB reset, LSB set"
        4,5,6,7) register[2:0]=0; "now restart the conversion"
```

```
      END;
END.


TEST_VECTORS
IN   clk, start, vo_gr_vi;
OUT  register[2:0];
BEGIN
"clk  start   vo_gr_vi   register[2:0]"
"-----------------------------------"
   p    x        x         L  L  L;   "Start with all reset"
   c    0        x         L  L  L;   "Stay reset since not started"
   c    1        x         L  H  L;   "Start and set MSB"
   c    x        0         L  H  H;   "keep MSB and set LSB"
   c    x        1         H  H  L;   "reset LSB and finish"
   c    x        x         L  L  L;   "go back to the start"
   c    1        x         L  H  L;   "start again"
   c    x        1         L  L  H;   "reset MSB set LSB"
   c    x        0         H  L  H;   "keep LSB and finish"
   c    x        x         L  L  L;   "go and start again"
END.


Boolean equations from Design [programmable_register]
register[2](H) = /register[2]*register[0] ;
register[2].enable = #1 ;
register[1](H) = start*/register[2]*/register[1]*/register[0] +
   /register[2]*register[1]*/vo_gr_vi +
   /register[2]*register[1]*register[0] ;
register[1].enable = #1 ;
register[0](H) = /register[2]*register[1]*/register[0] +
   /vo_gr_vi*/register[2]*register[0] ;
register[0].enable = #1 ;
```

Appendix 3 Semiconductor Manufacturers – Contact Addresses

The following is a selection of contact addresses for the manufacturers mentioned in the text. The UK address is given, except where this is inappropriate. There are of course more manufacturers than those listed. Also, as in any industry, change can be rapid indeed; takeovers and mergers continue apace – the address for Philips is poignant in this regard.

Analog Devices	Industrial Products Division, One Technology Way, P.O. Box 9106, Norwood, MA, USA
Advanced Micro Devices (UK) Ltd	Birkdale House, The Links, Kelvin Close, Birchwood, Warrington
Altera	21 Broadway, Maidenhead, Berks, SL6 1JK
Cypress Semiconductors	3901 North First St, San Jose, CA 95134, USA
GEC/ Plessey	Cheney Manor, Swindon, Wilts, SN2 2QW
Intel Corporation (UK) Ltd	Pipers Way, Swindon, Wilts, SN3 1RJ
Maxim Integrated Products Ltd	21C Horseshoe Park, Pangbourne, Reading, Berks, RG8 7JW
Motorola	European Literature Centre, 88 Tanners Drive, Blakelands, Milton Keynes, Bucks MK14 5BP
National Semiconductor (UK) Ltd	301 Harpur Centre, Horne Lane, Bedford, MK40 1TR
Philips Components Ltd	Mullard House, Torrington Place, London WC1E 7HD
Texas Instruments Ltd	Regional Technology Centre, Manton Lane, Bedford, MK41 7PA
Xilinx Inc.	2100 Logic Drive, San Jose, C.A., 95124, USA

Appendix 4 Further Reading

Abromovici, M., Breurer, M. A., and Friedman, A. D., *Digital Systems Testing and Testable Design* (Computer Science Press, New York, 1990)

Advanced Micro Devices, *PAL Device Databook*, AMD Inc., Birkdale House, The Links, Kelvin Close, Birchwood, Warrington

Altera, *Data Book*, Altera Inc., 21 Broadway, Maidenhead, Berks, SL6 1JK

Alvarez, A. R., *BICMOS Technology and Applications* (Kluwer Academic Publishers, Boston, 1989)

Analog Devices, *Data Converter Reference Manual* (vols. I and II), Analog Devices Inc.

Boole, G., *An Investigation into the Laws of Thought* (Dover Publications, New York, 1954)

Bolton, M., *Digital Systems Design with Programmable Logic* (Addison-Wesley, Wokingham, 1990)

Bostock, G., *Programmable Logic Handbook* (Butterworth-Heinemann, Oxford, 2nd edn, 1993)

Brayton, R. K., Hachtel, G. D., McMullen, C. T. and Sangiovanni-Vincentelli, A. L., *Logic Minimisation Algorithms for VLSI Synthesis* (Kluwer Academic Publishers, Boston, 1985)

Candy, J. C. and Temes, G. C., *Oversampling Delta Sigma Converters* (IEEE Press, New York, 1992)

Carroll, L., *Game of Logic* (Dover Publications, New York, 1958)

Catt, I., Walton, D. and Davidson, M., *Digital Hardware Design* (Macmillan, London, 1979)

Clare, C. R., *Designing Logic Systems Using State Machines* (McGraw-Hill, London, 1973)

Cypress, *CMOS Databook*, Cypress Semiconductors, 3901 North First St, San Jose, CA 95134, USA

Deitel, H. M. and Deitel, P. J., *How to Program* (Prentice-Hall, Englewood Cliffs, NJ, 1992)

Elmasry, M.I., *Digital Bipolar Integrated Circuits* (Wiley, Chichester, 1983)

Grebene, A. B., *Analog Integrated Circuit Design* (Van Nostrand-Reinhold, 1972)

Haskard, M. R., *An Introduction to Application Specific Integrated Circuits* (Prentice-Hall, 1990)

Haznedar, H., *Digital Microelectronics* (Benjamin Cummins, Redwood City, 1991)

Hayt, W. H. *Engineering Circuit Analysis* (McGraw-Hill, London, 5th edn, 1986)

Intel, *Programmable Logic*, Intel Corporation, Pipers Way, Swindon, Wilts, SN3 1RJ

Mead, C. and Conway, L., *Introduction to VLSI Systems* (Addison-Wesley, Reading, MA, 1980)

Meade, M. L. and Dillon, R., *Signals and Systems* (Chapman and Hall, 1991)

Millman, J. and Grabel, A., *Microelectronics* (McGraw-Hill, London, 2nd edn, 1988)

Muroga, S., *VLSI Systems Design* (Wiley, Chichester, 1982)

National Semiconductor, *Programmable Logic Devices Databook and Design Guide* and *Interface Databook*, National Semiconductor Inc., 301 Harpur Centre, Horne Lane, Bedford, MK40 1TR

Parker, G. J., *Introductory Semiconductor Device Physics* (Prentice-Hall, London, 1993)

Patterson, D. A. and Hennessy, J. L., *Computer and Organisation and Design – the Hardware/ Software Interface* (Morgan Kaufman, San Mateo, CA, 1993)

Philips, *Integrated Fuse Logic*, Philips Ltd, Mullard House, Torrington Place, London WC1E 7HD

Plessey, *Data Converters and Datacoms IC Handbook*, Plessey/GEC Ltd, Cheney Manor, Swindon, Wilts, SN2 2QW

Proakis, J. G., *Digital Communications* (McGraw-Hill, London, 2nd edn, 1989)

Prosser, D. and Winkel, F., *The Art of Digital Design* (Prentice-Hall, London, 2nd edn, 1989)

Rabiner, R. L. and Gold, B., *Theory and Application of Digital Signal Processing* (Prentice-Hall, London, 1975)

Sedra, A. and Smith, K., *Microelectronic Circuits* (Saunders, London, 3rd edn, 1991)

Seitzer, D., Pretzl, G. and Hamdy, N. A., *Electronic Analog to Digital Converters* (Wiley, Chichester, 1983)

Shaw, A.W., *Logic Circuit Design* (Saunders, London, 1993)

Stremler, F. G., *Introduction to Communication Systems* (Addison-Wesley, 3rd edn, London, 1990)

Swartzlander, E. E. (Ed.), *Computer Arithmetic*, vol. II (IEEE Computer Society Press, Los Alamitos, CA, 1990)

Texas Instruments, *TTL Handbook* (Texas Inc, 4 vols), *Linear Circuits Databook*, vol. 2, and *Interface Databook*, Texas Instruments Ltd, Regional Technology Centre, Manton Lane, Bedford, MK41 7PA

Thornton, E., *Electrical Interference and Protection* (Ellis Horwood, New York, 1990)

Unger, S. H., *The Essence of Logic Circuits* (Prentice-Hall, Englewood Cliffs, NJ, 1989)

Uyemura, J. P., *Fundamentals of MOS Integrated Circuits* (Addison-Wesley, New York, 1988)

Wakerly, J. F., *Digital Design Principles and Practices* (Prentice-Hall, London, 2nd edn, 1990)

Weste, N. and Estraghian, K., *CMOS VLSI Design* (Addison-Wesley, Reading, MA, 2nd edn, 1992)

Wilkinson, B. and Makki, R., *Digital System Design* (Prentice-Hall, London, 2nd edn, 1992)

Wilkins, B.R., *Testing Digital Circuits – An Introduction* (Van Nostrand-Reinhold, Wokingham, 1986)

Wiitala, S. A., *Discrete Mathematics: A Unified Approach* (McGraw-Hill, London, 1987)

Xilinx, *User Guide and Tutorials*, Xilinx Inc., 2100 Logic Drive, San Jose, C.A., 95124, USA

Appendix 5   Abbreviated Answers to Questions

Chapter 2

1 These can be shown using the laws of Boolean algebra.

2 By extracting two prime implicants from the K-map the solution is:

$$f = \overline{A} \cdot B \cdot \overline{D} + A \cdot \overline{B} \cdot D$$

This is a minimum result in terms of the specified gates. To implement the function using NOR gates we can either extract the '0's or just twice invert f.

3 For a hazard-free solution, by extracting four prime implicants from the K-map, one of which is a bridging term:

$$f = A \cdot \overline{C} + A \cdot \overline{D} + \overline{A} \cdot \overline{B} \cdot D + \overline{B} \cdot \overline{C} \cdot D$$

4 By using de Morgan's law and by expanding the EXOR term, then by using a K-map, we obtain a minimal solution as:

$$f = \overline{B} \cdot \overline{C} + B \cdot C$$

5 Again by applying de Morgan's law and by expanding the EXNOR terms, then by using a K-map we obtain:

$$f = \overline{A} \cdot \overline{C} + A \cdot C$$

6 The level of logic is three. The worst case propagation delay is 33 ns, pedantically 32 ns.

7 Attach a logic '1' input to channels 0, 2 and 6 (000, 010 and 110) and a logic '0' input to the remaining five.

8
```
f = ((NOT(A) AND NOT(B) AND C) OR (NOT(A) AND B AND NOT(C)) OR
     (A AND NOT(B) AND NOT(C)) OR (A AND B AND C))

IF ((NOT(A) AND NOT(B) AND C) OR (NOT(A) AND B AND NOT(C)) OR
     (A AND NOT(B) AND NOT(C)) OR (A AND B AND C)) THEN f=1
                                                   ELSE f=0
```

9 `CASE (A,B,C)`
　　`BEGIN`
　　　`000: f=0`
　　　`001: f=1`
　　　`010: f=1`
　　　`011: f=0`
　　　`100: f=1`
　　　`101: f=0`
　　　`110: f=0`
　　　`111: f=1`
　　`END`

10 The minimised result is

$$f = \overline{A} \cdot \overline{B} \cdot D + \overline{A} \cdot B \cdot \overline{D} + \overline{C}$$

For the algebraic solution, expand the EXOR terms, use de Morgan's law, and group common terms.

11 For I = driver, S = seat belt, D = door and A = safety,

$$f = I \cdot S \cdot D \cdot \overline{A}$$

To use NAND or NOR gates transform f using de Morgan's law:

12 The solutions for each function are

$$G = A \cdot \overline{C} + A \cdot B \cdot \overline{D} + B \cdot \overline{C} \cdot \overline{D}$$

$$E = \overline{A} \cdot \overline{B} \cdot \overline{C} \cdot \overline{D} + \overline{A} \cdot B \cdot \overline{C} \cdot D + A \cdot \overline{B} \cdot C \cdot \overline{D} + A \cdot B \cdot C \cdot D$$

$$L = \overline{A} \cdot C + \overline{A} \cdot \overline{B} \cdot D + \overline{B} \cdot C \cdot D$$

For NAND gates use de Morgan's law. For NOR extract the '0's. To implement L using an 8-1 multiplexer use control lines BCD addressing the eight input channels connected either to A or '1' or '0' covering all terms in L.

13 A static hazard can exist between the first two terms and the last two terms, these are where bridging terms can be used. The hazard-free output f_{hf} is given by

$$f_{hf} = \overline{A} \cdot C \cdot D + B \cdot C \cdot \overline{D} + \overline{A} \cdot B \cdot C + A \cdot \overline{C} \cdot D + A \cdot \overline{B} \cdot \overline{C}$$

The level of logic is 3, corresponding to a worst-case delay of 30 ns.

Chapter 3

1 One solution is to couple an inverter to the output of a 2-input NAND gate.

2 This can be achieved by including the inversion within the TTL NAND gate circuit, rather than at the output.

3 By inverting the inputs we will obtain the AND function.

4 N_{ML} rises to 1.1 V and N_{MH} falls by 0.5 V.

5 The load transistor's resistance must be twice that of the switching transistor.

6 A parallel connection of series NMOS transistors connected to GND gives a LOW output for $\overline{A \cdot B \cdot C \cdot D}$ and a series pair of parallel PMOS transistors connected to V_{DD} gives the HIGH output for $(\overline{A} + \overline{B}) \cdot (\overline{C} + \overline{D})$.

7 $R_1 = 400\,\Omega$ and $R_2 = 453\,\Omega$. The transistors do not reach saturation since V_{CE} is never sufficiently low.

Chapter 4

1 Use pull-down resistors (connected to 0 V rather than V_{CC}) on the bistable inputs and the single pole connection on the switch to $+V_{CC}$.

2 No, the D input changes within T_{SU}.

3 For a natural binary-encoded sequence of states where 111 is unused (going to 000), and for a coding $D_2 = \text{MSB}$ and $D_0 = \text{LSB}$

$$D_2 = Q_2 \cdot \overline{Q_1} + \overline{Q_2} \cdot Q_1 \cdot Q_0 \qquad D_1 = \overline{Q_1} \cdot Q_0 + \overline{Q_2} \cdot Q_1 \cdot \overline{Q_0} \qquad D_0 = \overline{Q_2} \cdot \overline{Q_0} + \overline{Q_1} \cdot \overline{Q_0}$$

4 The T inputs are $T_2 = \text{MSB}$ and $T_0 = \text{LSB}$:

$$T_0 = \overline{Q_2} + \overline{Q_1} \qquad T_1 = Q_0 + Q_2 \cdot Q_1 \qquad T_2 = Q_1 \cdot Q_0 + Q_2 \cdot Q_1$$

5 Use the ripple counter (Section 4.11) with CLR asynchronous inputs connected to

$$\overline{\text{CLR}} = \overline{Q_2 \cdot Q_1 \cdot Q_0}$$

6 The T-type solution and the D-type are identical. The unused state was not clearly specified for the T-types. The (simpler) asynchronous design enters the unused state and then resets leading to potential unreliability.

7 The asynchronous counter erroneously activates decoded outputs, so the synchronous counter would appear a wise choice.

```
8 CASE (Q[2..0])
    BEGIN
    #b000: BEGIN start=0
                    IF (go) THEN BEGIN Q[2..0]=#b000 END
                            ELSE BEGIN Q[2..0]=#b001 END
            END
    #b001: BEGIN start=0 Q[2..0]=#b010 END
    #b010: BEGIN start=1 Q[2..0]=#b011 END
    #b011: BEGIN start=1 Q[2..0]=#b100 END
    #b100: BEGIN start=0 Q[2..0]=#b000 END
    #b101,#b110, #b111: BEGIN start=0 Q[2..0]=#b010 END
```

9
```
IF (/Q[2]*/Q[1]*/Q[0]) THEN
                        BEGIN start=0
                            IF (go) THEN BEGIN Q[2..0]=#b000 END
                                    ELSE BEGIN Q[2..0]=#b001 END
                        END
    ELSE
    BEGIN IF (/Q[2]*/Q[1]*Q[0]) THEN BEGIN start=0
                                        Q[2..0]=#b010
                        END

        ELSE
        BEGIN IF (/Q[2]*Q[1]*/Q[0]) THEN BEGIN start=1
                                            Q[2..0]=#b011
                                        END
            ELSE
            BEGIN IF (/Q[2]*Q[1]*Q[0])  THEN BEGIN start=1
                                            Q[2..0]=#b100
                                        END
                ELSE
                BEGIN IF (Q[2]*/Q[1]*/Q[0]) THEN BEGIN start=0
                                                Q[2..0]=#b000
                                            END
                    ELSE
                    BEGIN  Start=0
                            Q[2..0]=#b000
                    END
                END
            END
        END
    END
END
```

10 The overall design is functionally equivalent to multiplexer-selecting an output from a set of inputs in sequence controlled by an up/down counter. Using a modulo 6 counter and decoding the outputs ($Q_2Q_1Q_0 = 000$ to select A, $Q_2Q_1Q_0 = 001$ and 011 to select B, $Q_2Q_1Q_0 = 100$ and 110 to select C, $Q_2Q_1Q_0 = 101$ to select D) where the sequence is $Q_2Q_1Q_0 = 000, 001, 100, 101, 110, 011$ with states 010 and 111 unused (returning to 000), the solution is

$$D_0 = \overline{Q}_1 \cdot \overline{Q}_0 + Q_2 \cdot \overline{Q}_0$$
$$D_1 = Q_2 \cdot \overline{Q}_1 \cdot Q_0 + Q_2 \cdot Q_1 \cdot \overline{Q}_0$$
$$D_2 = \overline{Q}_1 \cdot Q_0 + Q_2 \cdot \overline{Q}_1$$

and

$$\text{Select A} = \overline{Q}_2 \cdot \overline{Q}_0$$
$$\text{Select B} = \overline{Q}_2 \cdot Q_0$$
$$\text{Select C} = Q_2 \cdot \overline{Q}_0$$
$$\text{Select D} = Q_2 \cdot Q_0$$

11 For Q_0 latching the input data and Q_1 latching Q_0 (in the shift register) and for a 2-bit counter which counts in natural binary we have

$$D_1 = \overline{Q}_1 \cdot Q_0 + Q_0 \cdot S/\overline{C} + Q_1 \cdot \overline{Q}_0 \cdot \overline{S/\overline{C}}$$
$$D_0 = S/\overline{C} \cdot \text{data} + \overline{Q}_0 \cdot \overline{S/\overline{C}}$$

The max frequency $= 21.3\,\text{MHz}$

12 Use a 3-bit bidirectional shift register (Section 4.8.4) and a modulo 3 counter to clock it. A natural binary modulo 3 counter is

$$D_1 = \overline{Q}_1 \cdot Q_0$$
$$D_0 = \overline{Q}_1 \cdot \overline{Q}_0$$

Chapter 5

1
```
CASE (A,B,C)
  BEGIN   ;enable specified line, disable others
  #b000: BEGIN line_1=1 line_2=0 line_3=0 line_4=0
                line_5=0 line_6=0 line_7=0 line_8=0 END
  #b001: BEGIN line_1=0 line_2=1 line_3=0 line_4=0
                line_5=0 line_6=0 line_7=0 line_8=0 END
  #b010: BEGIN line_1=0 line_2=0 line_3=1 line_4=0
                line_5=0 line_6=0 line_7=0 line_8=0 END
  #b011: BEGIN line_1=0 line_2=0 line_3=0 line_4=1
                line_5=0 line_6=0 line_7=0 line_8=0 END
  #b100: BEGIN line_1=0 line_2=0 line_3=0 line_4=0
                line_5=1 line_6=0 line_7=0 line_8=0 END
  #b101: BEGIN line_1=0 line_2=0 line_3=0 line_4=0
                line_5=0 line_6=1 line_7=0 line_8=0 END
  #b110: BEGIN line_1=0 line_2=0 line_3=0 line_4=0
                line_5=0 line_6=0 line_7=1 line_8=0 END
  #b111: BEGIN line_1=0 line_2=0 line_3=0 line_4=0
                line_5=0 line_6=0 line_7=0 line_8=1 END
  END
```

with solution

```
LINE_8  =  C * B * A
LINE_7  =  /C * B * A
LINE_6  =  C * /B * A
LINE_5  =  /C * /B * A
LINE_4  =  C * B * /A
LINE_3  =  /C * B * /A
LINE_2  =  C * /B * /A
LINE_1  =  /C * /B * /A
```

2
```
CASE (A,B,C)
  BEGIN
  #b000: BEGIN output = line_1 END
  #b001: BEGIN output = line_2 END
  #b010: BEGIN output = line_3 END
  #b011: BEGIN output = line_4 END
  #b100: BEGIN output = line_5 END
  #b101: BEGIN output = line_6 END
  #b110: BEGIN output = line_7 END
  #b111: BEGIN output = line_8 END
  END
```

with solution showing exactly how the multiplexer includes a decoder (as given in the solution to 5.1)) as

```
OUTPUT = LINE_8 * C * B * A + /C * B * A * LINE_7
       + C */B * A * LINE_6 + /C * /B * A * LINE_5
       + C * B * /A * LINE_4 + /C * B * /A * LINE_3
       +  C * /B * /A * LINE_2 +  /C * /B * /A * LINE_1
```

3

```
IF (/A*B*/C*/D+/A*B*C*/D+A*/B*/C*D+A*/B*C*D)
                                          THEN BEGIN f=1   END
                                          ELSE BEGIN f=0 END
```

with solution   F = A * /B * D + /A * B * /D

4

```
f = /B*/C*/D+/B*/C*/(A*D)+A*/B*/C+B*C*(A:+:D)+/A*B*C+B*C*/(/A*/D)
```
The solution is   F = B * C + /B * /C

5

```
IF ((/A*/B*C) +(/A*B*/C) + (A*/B*/C) + (A*B*C))
                                 THEN BEGIN f=1 END
                                 ELSE BEGIN f=0 END
```

This cannot be minimised, since no further reduction is possible. The solution via a CASE statement is

```
CASE (A,B,C)
    BEGIN
    #b000: BEGIN f=0 END
    #b001: BEGIN f=1 END
    #b010: BEGIN f=1 END
    #b011: BEGIN f=0 END
    #b100: BEGIN f=1 END
    #b101: BEGIN f=0 END
    #b110: BEGIN f=0 END
    #b111: BEGIN f=1 END
    END
```

This could clearly be reduced using the OTHERWISE construct in PALASM. The solution, expressed in terms of /F, is

```
/F  =  /C * B * A  +  C * /B * A  +  C * B * /A  +  /C * /B *
/A
```

6
```
CASE (q[1..0])
   BEGIN ;define sequence of next states
      #b00: BEGIN q[1..0]=#b01 END ;natural binary sequence
      #b01: BEGIN q[1..0]=#b10 END
      #b10: BEGIN q[1..0]=#b00 END
      #b11: BEGIN q[1..0]=#b00 END ;unused state to start
   END
q[1..0].CLKF = clk
```

with the solution for a modulo 3 counter as

```
Q[0]   :=  /Q[0] * /Q[1]      Q[0].CLKF  =  CLK
Q[1]   :=  Q[0] * /Q[1]       Q[1].CLKF  =  CLK
```

7
```
IF model_r THEN BEGIN q[2] = q[1]
                       q[1] = data_in_lr END
             ELSE BEGIN q[2] = data_in_rl
                       q[1] = q[2] END
q[2..1].CLKF = clk
```

with solution:

```
Q[1]   :=  Q[2] * /MODEL_R   +  MODEL_R * DATA_IN_LR
   Q[1].CLKF  =  CLK
   Q[2]   :=  DATA_IN_RL * /MODEL_R  +  MODEL_R * Q[1]
   Q[2].CLKF  =  CLK
```

8
```
CASE (Q[2..0])
   BEGIN
   #b000: BEGIN start=0
                 IF (go) THEN BEGIN Q[2..0]=#b000 END
                         ELSE BEGIN Q[2..0]=#b001 END
          END
   #b001: BEGIN start=0
                 Q[2..0]=#b010
          END
   #b010: BEGIN start=1
                 Q[2..0]=#b011
          END
```

```
#b011: BEGIN start=1
                Q[2..0]=#b100
       END
#b100: BEGIN start=0
                Q[2..0]=#b000
       END
OTHERWISE: BEGIN start=0
                   Q[2..0]=#b000
              END
END
Q[2..0].CLKF = clk
```

with solution

```
START   =  Q[1] * /Q[2]
Q[0]    := /Q[0] * Q[1] * /Q[2]   +  /Q[0] * /Q[2] * /GO
Q[0].CLKF  =  CLK
Q[1]    := /Q[0] * Q[1] * /Q[2]   +   Q[0] * /Q[1] * /Q[2]
Q[1].CLKF  =  CLK
Q[2]    :=  Q[0] * Q[1] * /Q[2]
Q[2].CLKF  =  CLK
```

The complete solution to Question 9 expressed in PALASM is

```
IF (/Q[2]*/Q[1]*/Q[0]) THEN
                          BEGIN start=0
                                IF (go) THEN BEGIN Q[2..0]=#b000 END
                                        ELSE BEGIN Q[2..0]=#b001 END
                          END
ELSE
BEGIN IF (/Q[2]*/Q[1]*Q[0]) THEN BEGIN start=0
                                       Q[2..0]=#b010
                                 END

  ELSE
  BEGIN IF (/Q[2]*Q[1]*/Q[0]) THEN BEGIN start=1
                                         Q[2..0]=#b011
                                   END

        ELSE
        BEGIN IF (/Q[2]*Q[1]*Q[0]) THEN BEGIN start=1
                                             Q[2..0]=#b100
                                       END

              ELSE
              BEGIN IF (Q[2]*/Q[1]*/Q[0]) THEN BEGIN start=0
                                                    Q[2..0]=#b000
                                              END
```

```
                                        ELSE
                                        BEGIN  Start=0
                                               Q[2..0]=#b000
                                        END
                                END
                          END
                   END
            END
      END
      Q[2..0].CLKF = clk
```

9

```
      CASE (A,B,C,D)
             BEGIN    ;specify G E and L
             #b0000: BEGIN G=0 E=1 L=0 END
             #b0001: BEGIN G=0 E=0 L=1 END
             #b0010: BEGIN G=0 E=0 L=1 END
             #b0011: BEGIN G=0 E=0 L=1 END
             #b0100: BEGIN G=1 E=0 L=0 END
             #b0101: BEGIN G=0 E=1 L=0 END
             #b0110: BEGIN G=0 E=0 L=1 END
             #b0111: BEGIN G=0 E=0 L=1 END
             #b1000: BEGIN G=1 E=0 L=0 END
             #b1001: BEGIN G=1 E=0 L=0 END
             #b1010: BEGIN G=0 E=1 L=0 END
             #b1011: BEGIN G=0 E=0 L=1 END
             #b1100: BEGIN G=1 E=0 L=0 END
             #b1101: BEGIN G=1 E=0 L=0 END
             #b1110: BEGIN G=1 E=0 L=0 END
             #b1111: BEGIN G=0 E=1 L=0 END
             END
```

with solution

```
      L  =  D * C * /B   +   C * /A   +   D * /B * /A
      G  =  /C * A  +   /D * /C * B   +   /D * B * A
      E  =  D * C * B * A  +   /D * C * /B * A
         +  D * /C * B * /A   +   /D * /C * /B * /A
```

10

```
      CASE (q[2..0])
        BEGIN ;define sequence of next states
           #b000: BEGIN q[2..0]=#b001 END
           #b001: BEGIN q[2..0]=#b010 END
           #b010: BEGIN q[2..0]=#b011 END
           #b011: BEGIN q[2..0]=#b100 END
```

```
      #b100: BEGIN q[2..0]=#b101 END
      #b101: BEGIN q[2..0]=#b000 END
      OTHERWISE: BEGIN q[2..0]=#b000 END
   END
q[2..0].CLKF = clk
```

with solution

```
    Q[0]   :=  /Q[0] * /Q[1]  +  /Q[0] * /Q[2]
    Q[0].CLKF  =  CLK
    Q[1]   :=  /Q[0] * Q[1] * /Q[2]   +  Q[0] * /Q[1] * /Q[2]
    Q[1].CLKF  =  CLK
    Q[2]   :=  /Q[0] * /Q[1] * Q[2]   +  Q[0] * Q[1] * /Q[2]
    Q[2].CLKF  =  CLK
```

This can be reduced by a grey code implementation as

```
CASE (q[2..0])
  BEGIN ;define sequence of next states
    #b000: BEGIN q[2..0]=#b001 END
    #b001: BEGIN q[2..0]=#b011 END
    #b011: BEGIN q[2..0]=#b111 END
    #b111: BEGIN q[2..0]=#b110 END
    #b110: BEGIN q[2..0]=#b100 END
    #b100: BEGIN q[2..0]=#b000 END
    #b010: BEGIN q[2..0]=#b000 END   ;allocate unused states
    #b101: BEGIN q[2..0]=#b100 END   ;using grey code approach
  END
q[2..0].CLKF = clk
```

with a more compact solution

```
    Q[0]   :=  Q[0] * /Q[2]   +  /Q[1] * /Q[2]
    Q[0].CLKF  =  CLK
    /Q[1]  :=  /Q[0]   +  /Q[1] * Q[2]
    Q[1].CLKF  =  CLK
    /Q[2]  :=  /Q[0] * /Q[1]  +  /Q[0] * /Q[2]  +  /Q[1] * /Q[2]
    Q[2].CLKF  =  CLK
```

11

```
CASE (q[1..0])
  BEGIN ;define sequence of next states
  #b00: BEGIN IF shift_count THEN BEGIN
                          IF data THEN BEGIN q[1..0]=#b01 END
                          ELSE BEGIN q[1..0]=#b00 END
```

```
                                        END
                         ELSE BEGIN q[1..0]=#b01 END
           END
     #b01: BEGIN IF shift_count THEN BEGIN
                            IF data THEN BEGIN q[1..0]=#b11 END
                                     ELSE BEGIN q[1..0]=#b10 END
                         END
                         ELSE BEGIN q[1..0]=#b10 END
           END
     #b10: BEGIN IF shift_count THEN BEGIN
                            IF data THEN BEGIN q[1..0]=#b01 END
                                     ELSE BEGIN q[1..0]=#b00 END
                         END
                         ELSE BEGIN q[1..0]=#b11 END
           END
     #b11: BEGIN IF shift_count THEN BEGIN
                            IF data THEN BEGIN q[1..0]=#b11 END
                                     ELSE BEGIN q[1..0]=#b10 END
                         END
                         ELSE BEGIN q[1..0]=#b00 END
           END
     END
q[1..0].CLKF = clk
```

with solution

```
Q[0]   := DATA * SHIFT_COUNT + /SHIFT_COUNT * /Q[0]
Q[0].CLKF  =  CLK
/Q[1]  := /Q[0] * /Q[1] + SHIFT_COUNT * /Q[0]
     + /SHIFT_COUNT * Q[0] * Q[1]
Q[1].CLKF  =  CLK
```

Chapter 6

1   The solution for bits D, E and F converted from bits A, B and C is

$$F = B \oplus C \qquad E = A \oplus B \qquad D = A$$

2   The extraction is then, via PALASM,

```
Z  =  /D * /C * B  +  /D * /C * A   +  D * C * B * /A
Y  =  /D * B  +  /D * /C * A   +  D * C * /B * A
/X  =  /B * /A  +  D * /B  +  C * /A
W  =  /C * A  +  /D * /C * B
```

3 and 4   The result is the same as in Section 6.3.2.2.

5   Use a truth table for the memory and a circuit implementing

```
IF (input<=5) THEN output:=input+3 ELSE output:=2*input+3;
```

using a digital comparator to give a signal for 2-1 line multiplexers to choose either the input bits or the input bits shifted left one place (equivalent to multiplication by two), followed by an adder adding three. To implement it using a PAL we need to program the PAL using a CASE construct expressing the truth table.

6   (a)   Use a 2-1 line multiplexer controlled by the mode signal M to choose either the output of the adder (half adder plus full adder) or the subtracter (half subtracter plus full subtracter).
    (b)   Use the mode signal to gate the complemented form of B to a set of full adders and also to set the carry input to the LSB as '1'.
    (c) B_0 is clearly the same for both circuits.

Chapter 7

1   A successive approximation converter will suffice.

2   An 8-bit converter.

3   Only the counter part can be accommodated in a PAL.

4   Use a counter counting down and a DAC feeding the output back to a comparator enabling the counter to count down while $V_o > V_i$, and to stop when $V_o \leq V_i$.

5   The MSB will have undue weight.

6   For a PAL

```
;register coding is regster[2:1:0] = [conversion complete:MSB:LSB]
CASE (regster[2..0])
BEGIN;first test to see whether we start or not
7: BEGIN IF (start) THEN
                BEGIN regster[2..0]=5 END;if start then reset MSB
                ELSE
                BEGIN regster[2..0]=7 END;otherwise stay here
     END
5: BEGIN IF (vo_gr_vi) THEN
                BEGIN regster[2..0]=4 END;reset MSB, reset LSB
                ELSE
```

```
                                  BEGIN regster[2..0]=6 END;keep MSB, reset LSB
         END
4: BEGIN IF (vo_gr_vi) THEN
                                  BEGIN regster[2..0]=0 END;reset MSB, reset LSB
                                  ELSE
                                  BEGIN regster[2..0]=1 END;reset MSB, set LSB
         END
6: BEGIN IF (vo_gr_vi) THEN
                                  BEGIN regster[2..0]=2 END;set MSB, reset LSB
                                  ELSE
                                  BEGIN regster[2..0]=3 END;set MSB, set LSB
         END
OTHERWISE: BEGIN regster[2..0]=7 END;now restart the conversion
END
```

The difference in function is that the designs will approach a solution in different ways.

7   Use a 3-bit converter with an appropriate priority encoder.

Index